Smokestacks and Progressives

Smokestacks and Progressives

Environmentalists, Engineers, and Air Quality in America, 1881–1951

David Stradling

The Johns Hopkins University Press

Baltimore and London

© 1999 The Johns Hopkins University Press
All rights reserved. Published 1999
Printed in the United States of America on acid-free paper

9 8 7 6 5 4 3 2 1

The Johns Hopkins University Press
2715 North Charles Street
Baltimore, Maryland 21218-4363
www.press.jhu.edu

Library of Congress Cataloging-in-Publication Data will be found
at the end of this book.
A catalog record for this book is available from the British Library.

ISBN 0-8018-6083-0

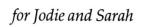

for Jodie and Sarah

Contents

Acknowledgments

Although he does not remember it, this project began in the office of Zane Miller, at the University of Cincinnati. I was a Wisconsin graduate student at the time, working with Zane's friend Stanley Schultz. When I told Zane I wanted to research some aspect of Cincinnati's environmental history, he led me away from the sewers and toward the skies and smoke. I am forever indebted to Zane for that initial guidance, for he pointed me down the very productive and deeply rewarding path I have followed for close to a decade.

During the past eight years, I have relied on the kindness and expertise of numberless librarians. I received generous help at the many excellent libraries on the campus of the University of Wisconsin–Madison. I am also beholden to librarians at the Cincinnati Historical Society, the University of Cincinnati Library, the Hamilton County Public Library, the Historical Society of Western Pennsylvania, the Carnegie Library of Pittsburgh, the Mellon Institute, the University of Pittsburgh Library, the New York City Central Research Library, and the very efficient Milwaukee Municipal Reference Library. I am especially grateful for an extremely productive stay at the Hagley Library, in Wilmington, Delaware, a facility with superbly kept archives and generous financial aid for scholars. For two critical years I also relied on David Hughes and the rest of the Colgate University library staff, all of whom were generous with their time and help.

In the course of this work I was blessed with two very skilled and enthusiastic advisors. Zane Miller, who, after getting this project started in 1990, served as my doctoral advisor for one very productive year, seemed at times more interested in my work than I was myself. Some days I received multiple "Zane-o-grams," memos with his latest thoughts on smoke. Back in Madison, Bill Cronon provided ample intellectual guidance and psychological support for four years as I completed this project. Bill edits as well as he writes, and he greatly improved my work. Just as important, Bill is a wonderful teacher, and the years I spent in the community of environmental scholars that he has created at Madison were irreplaceable. Others at the University of Wisconsin also aided in my work, including Stan Schultz, Colleen Dunlavy, Diane Lindstrom, Art McEvoy, and John Cooper. I would also like to thank my Madison colleague Peter Thorsheim, with whom I spent long hours discussing smoke. Jon Rees and Paul Taillon also shared ideas and information. I also benefitted from the comments of my fellow Frontiers in Urban Research seminarians at the University of Cincinnati.

During the process of turning my dissertation into a book, I received important advice from Joel Tarr, who read the manuscript in its entirety—twice. Joel's insights into the history of pollution control and fuel transitions proved invaluable to me in the late stages of the process, and I owe him the deepest of intellectual debts. This work also benefited from a remarkably close reading by Phil Scranton, to whom I am grateful. Although I have received considerable help along the way, as these acknowledgments should make clear, any errors and weaknesses that remain in the book are my responsibility alone.

This book is dedicated to my wife Jodie, who knows too well how much time and energy this project has demanded over the last seven years, and to our new daughter Sarah, who will only know this work as a book on the shelf—with her name up front.

Smokestacks and Progressives

Vision Obscured

In May of 1899, a New York *Tribune* editorial mourned the costs of several recent environmental disasters. The piece, entitled "When the Milk Is Spilled," recounted the gradual destruction of New York City's small parks, the reckless demolition of the nation's forests, and the rapid pollution of its waterways. "Once brooks and rivers were pure and sweet," the editorial noted, "but men made them receptacles for sewage until they became unutterably foul." The author lamented, "How vastly better if the parks had been preserved, and the shade trees protected, and the forests maintained, and the streams kept pure!"[1]

With the damage already done to the urban parks and the nation's forests and waterways, the *Tribune* predicted that coal smoke would produce the next great environmental disaster. While many other industrial cities, Pittsburgh, Chicago, and Cincinnati among them, had long experienced dense smoke pollution, in the 1890s New York remained renowned for its pure air. But in that decade some businesses began to turn to highly polluting soft coal for power. As the editorial asserted, "men are deliberately, wantonly polluting [the atmosphere] and bringing it into the smoky, murky, suffocating state known hitherto to less favored cities." The author feared that the people would fail to act before it was too late, as had been the case with the other environmental disasters of the late 1800s. "And one of these days," the author mourned, "when the mischief is fully done, when our once pellucid and crystalline atmosphere is transformed into Chicago reek, and Pittsburgh smoke, and London fog, men will begin to realize what they have lost, and will hold conventions, and pass resolutions, and enact laws, and spend great sums of money for the undoing of the mischief and the restoration of our atmosphere to its original state."[2]

The author could hardly have been more prophetic. In the next seventeen years middle-class reformers, not just in New York but in dozens of industrial cities, would realize what they had lost to smoke and would hold conventions and pass resolutions, enact laws, and spend great sums of money in the attempt to purify their air. This editorial was more than prophetic, however. Taken in total, with its concerns for past and future environmental degradation, the essay reflected the growing appreciation of the need to take action to preserve the beauty, healthfulness, and cleanliness of the environment in the 1890s. Indeed, this editorial reflected the growth of an environmentalist sentiment in turn-of-the-century industrial cities.[3]

Through the progressive decades, from the 1890s through the 1910s, secure middle-class urbanites reflected on the cities they had in large part constructed. Often from suburban residences, they looked upon badly flawed creations, and yet most felt confident that modest changes—destroying tenement districts, improving waste collection, controlling prostitution, and reducing smoke, among others—could render their cities moral and efficient. The irony, of course, was that the social structure that had placed middle-class reformers in a position to improve their cities had also caused or exacerbated the many problems that required solutions. This was particularly true of smoke. Most of the antismoke activists had benefitted personally from the urban-industrial order, profiting from the economic growth that had fueled the air pollution problem. So closely connected were smoke and economic growth in the minds of urban Americans of all classes that smoke symbolized prosperity, and images of thick smoke, both literary and pictorial, frequently represented economic health in turn-of-the-century America.

Progressive reformers, including antismoke activists, rarely offered a comprehensive critique of the industrial order that lay at the root of the diverse problems they hoped to solve. Some reformers did organize against specific industries and even specific companies, but for most progressives the object of reform was to preserve the industrial system that had so enriched their communities and themselves. In the decades before World War I, most urban reformers did not pine for an agrarian past; they worked for a better urban future. Industry and commerce had built a new civilization, and the American faith in progress prevented widespread yearning for retreat.[4]

The key to reform, then, lay in the ability of prominent citizens to offer a dominant definition of civilization, of what it actually should look like. By controlling the civilization rhetoric, middle- and upper-class urbanites could determine which aspects of urban America required reform, such as prostitution and smoke, and which aspects required none, such as the sanctity of private property and private profit. By the last decade of the 1800s, the notion of a clean, healthful, and attractive urban environment had become an important part of the Victorian conception of civilized life. Thus the environmental reformism that played such an important part in progressive activism of the next decades had its roots in the middle-class ideal of creating a more civilized society.[5]

Unfortunately for those who would reform the urban environment, the idea of civilization was as complex and contradictory as modern cities themselves. For as surely as filth and pollution could symbolize the uncivilized, so too could smoke symbolize the height of industrial civilization. An 1896 Republican campaign poster revealed the intimate connection between smoke and civilization, as William McKinley stood holding an American flag before the slogan "Prosperity at Home, Prestige Abroad." At his sides lay "Commerce," represented by tall-masted ships, and "Civilization," represented by factories belching black smoke. The word "Civilization" could not have been closer to the pollution unless the smoke streams themselves had spelled out the letters. Of course, for many Gilded Age urbanites smoke did spell civilization, and they saw it daily in their skies as clearly as they could in the poster.[6]

Smoke could not remain just a symbol, however, any more than it could remain aloft. As smoke hung over the city in dense clouds and fell in the form of black, acidic soot, it posed a multitude of serious problems for industrial cities. The dark clouds obstructed vistas and blocked sunlight, adding gloom to the city and often necessitating the use of artificial light even in daylight hours. The soot, much more troublesome than the smoke itself, helped make industrial cities filthy, inside and out, as it clung to clothes, furniture, drapes, and rugs. Soot darkened buildings and its acids ate into stone and steel alike. Merchants complained of unsold goods spoiling on store shelves, housewives protested against never-ending cleaning, and doctors warned that urban lungs were harmed by smoky air.

The problem for those who would reform the smoky environment became, then, how to use the troublesome reality of smoke to destroy the image of smoke as a sign of progress without threatening the idea of progress itself. How could middle-class urbanites offer a viable critique of smoke pollution, and other industrial pollution for that matter, without seeming to critique the industrial order? Could reform-minded citizens define smoke as a problem while continuing to exalt the industrial productivity that caused smoke?

Many of the earliest antismoke activists were middle- and upper-class women steeped in Victorian ideas of cleanliness, health, aesthetics, and morality. Their sense of environmental reform grew from the relationship they thought existed between cleanliness and beauty, as well as that between cleanliness and the general physical and moral health of human beings. In attempting to abate coal smoke, then, these women, and other reformers influenced by such Victorian ideas, created an environmental philosophy that aligned beauty, health, and cleanliness with prosperity and security as worthy goals of civilization. This philosophy was the necessary foundation of the antismoke crusade and other movements to improve the urban environment. Concerned with their urban habitat, many city residents organized, researched, publicized, lobbied, and even filed suits in an attempt to improve their environment. The rhetoric of the early smoke abatement crusade, focusing primarily on health and beauty, reveals that the philosophy driving the antismoke movement was closely akin to modern environmentalism.[7]

Most discussion of the development of environmentalism has focused on the affluence of the post–World War II era, concluding that wealth allowed suburban residents to seek environmental amenities in the late twentieth century. But neither the wealth nor the environmental degradation of the postwar era were new to American cities. Concern for health, aesthetics, and recreational space, the very same concerns that drove the postwar environmental movement, also propelled environmental reformers decades earlier. Indeed, the late 1800s and the early 1900s contain abundant examples of urban and suburban environmental activism, much of it successful. While some historians have produced excellent work concerning these efforts, most surveys of the progressive era still largely ignore them.[8] The prosperity of postwar suburbs may explain the rapid growth of recent environmental activism, but prosperity alone cannot explain the movement's true origins. After achieving some affluence, why did American suburbanites search for *environmental* amenities? In other words, postwar environmentalists had to develop both the means and the interest in improving their environments. The interest derived from an environmentalist philosophy developed decades earlier, in the middle-class Victorian values of late-nineteenth-century American cities.

The story of the antismoke movement complicates not only the argument concerning the postwar origins of environmentalism, but also the definition of conservationism. Generally historians have defined conservationism as a progressive movement to improve the nation's use of important natural resources, especially water and timber. Most conservation histories do not look far beyond the rivers and forests of the American West. While in most histories conservationism remains almost exclusively a western, rural, and sylvan movement, in actuality it entailed much more than the protection of western watersheds through government management of forests and grasslands.[9] It was a broad-based crusade designed to manage efficiently all of the nation's resources, both in public and in private hands. The rhetoric that dominated the later antismoke movement, with its emphasis on the promotion of efficiency and the conservation of coal through smoke reduction, serves as an obvious example of how conservationism penetrated progressive cities. Indeed, conservation included a myriad of movements in cities to conserve resources, health, and beauty. Conservationism inhabited progressive cities as well as western lands.

What follows is the story of the transformation of an early environmental movement. The antismoke crusade consisted of three phases. The early phase, begun in the 1890s, deeply reflected middle-class Victorian ideals, particularly the desire for the creation of a higher civilization through the development of a beautiful and moral environment. Out of these Victorian ideals, antismoke reformers led a crusade to purify their urban air. The second phase, beginning in the early 1910s, reflected the influence of the middle class's growing faith in science, technology, expertise, and economic progress. Out of these progressive ideals, reformers created a conservationist movement to increase efficiency and economy by decreasing the waste associated with smoke. Finally, in the late 1930s, after years of negligible progress, the movement shifted again, with the appearance of a new tactic designed to purify fuel rather than improve combustion. This last phase arrived even as the industrial nation moved away from its nearly complete dependence on coal; the imposition of fuel restrictions reflected the new weakness of the coal industry. The earlier environmental reformism and the conservationist arguments had failed to eliminate the smoke. With coal's loss of influence within the American economy, and its declining contribution to the nation's air pollution problem, the smoke clouds that had hung over industrial cities for decades dissipated in the 1940s and 1950s.

one

Coal

The Vital Essence of Our Civilization

But coal is in a sense the vital essence of our civilization. If it can be preserved, if the life of the mines can be extended, if by preventing waste there can be more coal left in this country after we of this generation have made every needed use of this source of power, then we shall have deserved well of our descendants.
—Gifford Pinchot, *The Fight for Conservation*, 1910

Civilized—Reclaimed from savage life and manners; instructed in arts, learning, and civil manners; refined; cultivated.
—*Webster's American Dictionary*, 1880

When in May of 1902, 145,000 anthracite coal miners began a strike in the hills of eastern Pennsylvania, the residents of East Coast cities took immediate and uncommon interest. For the second time in three years a United Mine Workers walkout paralyzed production in the anthracite district. The strike and its implications became front page news in New York City and remained so for six months. Just five days after mining ceased, the New York *Times* warned "Long Strike May Mean A Coal Famine Here." The cost of anthracite quickly jumped from $5.35 to $6.35 per ton in the city, on its way up to the $16 per ton it would reach before the strike ended. Many anthracite users promptly turned to bituminous coal, a cheaper but dirtier alternative, shipped from the more distant but still-operating mines of western Pennsylvania and West Virginia.[1]

When the *Times* warned of an impending "coal famine" in the early days of the anthracite strike, the choice of words could not have been more appropriate. The city would quite literally have starved without coal. By 1902 the factories of the East and the Midwest had long relied on coal for power and heat. One early observer reveled, "Coal is to the world of industry what the sun is to the natural world, the great source of light and heat with their innumerable benefits."[2] But coal did more than fuel the factories of the city; it did more than keep the city at work. Coal, as used by railroads, tugboats, and steamships, moved goods and people into and out of the city, and as used by elevated railways and ferries, coal moved people and goods within the city. In elevators, too, coal moved people and goods inside the city's sky-scrapers. Just as important, the heat supplied by coal fires kept the city warm at home and at work. Even in the hottest months of the year, as the anthracite strike began, New Yorkers could dread the coming of a winter without coal. And New Yorkers were not the only Americans who feared the shortage of coal in winter, as coal supplied heat for urbanites throughout the Midwest, Mid-Atlantic, and Northeast. So complete was this reliance on coal that one man wrote in 1887, "It is doubtful if we could even live more than a few weeks were our supplies of fuel suddenly cut off. The northern states would freeze and every state would starve."[3]

The story of smoke cannot be told without considerable attention to coal, for no one could think of smoke without thinking of the fuel that caused it. And few urbanites underestimated the value of coal to their civilization. Coal pervaded every sector of America's industrial cities. Every class of resident saw it, handled it, purchased it, and smelled its dust. Residents knew good coal from bad and which coals burned best in their furnaces. They knew the names of mines and mining regions that labeled the black diamonds they produced. Middle-class homeowners bought coal by the ton and stored it in basement bins. The working poor bought it by the bucket and used it spar-ingly in their tenement stoves for heating and cooking. The desperately poor dug through ash dumps in search of bits of unburned coal or combed the rail-road tracks for lumps that had fallen from rail cars that moved haltingly across city streets.[4]

Just as the soot from smoke and the dust from unburned coal mingled on city surfaces, so too did the issues of smoke abatement and coal use unite in the minds of city residents. Just as smoke and soot pervaded the city, so too did coal. And while almost all city residents had significant contact with coal, at the turn of the century almost one in every forty workers in the nation, nearly a million men, made their living digging coal from the earth or shov-eling it into fires. In 1910, on an average day over 725,000 men worked the nation's coal mines, most of them underground. Thousands of others did not make the United States census lists of mine workers, because they were too young for official calculation. Boys as young as ten worked the breakers at the mine mouths, where they picked bits of slate from the coal as it passed beneath them on conveyors. In 1910, the census also reported that over

100,000 men made a living tending coal fires. In Chicago alone 3,057 men shoveled coal into fires for a living. In New York City that figure stood at 7,320. Over 75,000 others fed coal into locomotive fireboxes around the nation. In addition, thousands of men made a living retailing and wholesaling coal in cities, and in turn they employed thousands of others to labor in their coal yards. Taken in total, including those who dug it, those who shipped it, those who sold it, and those who shoveled it into fires, coal directly employed hundreds of thousands of men of all classes.[5]

In addition to those who made their living in the coal industry, many large corporations depended on coal for more than just an inexpensive fuel supply. Several railroad companies relied on coal for a significant portion of their income. The East's most important lines carried considerable amounts of the fuel, including the Baltimore and Ohio, the Pennsylvania, the Chesapeake and Ohio, and the Reading Railroads. Coal became so important to some railroad companies that they purchased land in the Pennsylvania and West Virginia coal fields to ensure supply and profits. In and of itself, then, the coal industry formed an important sector in the American economy.[6]

Even as the middle-class reformers who worked for smoke abatement began their crusade in the late 1800s, they well understood the importance of coal to their cities and the relationship between air pollution and coal markets. Indeed, the composition of local coal supplies largely determined the relative smokiness of cities and influenced perceptions of the smoke. American markets offered two very different types of coal: bituminous (soft) and anthracite (hard).[7] Both were the carbonaceous remains of ancient plants, stored solar energy, in the form of dark black veins under rock and earth. The two types of coal differed significantly in their chemical compositions and, consequently, in the amount of smoke they produced. Bituminous coal contained a much higher percentage of volatile matter, such as water, hydrocarbons, sulphur, and other "impurities," all of which could be driven off under low heat, leaving behind the carbon which constituted the majority of the coal. In fact, the greater quantity of volatile matter in bituminous coal made it ideal for the creation of coke. When heated in controlled conditions, without oxygen, bituminous coal yielded its volatile matter, leaving behind a purer carbon fuel that burned hotter and cleaner than coal. This fuel, coke, while more expensive than unprocessed coal, found a ready market among steel producers who valued its pure, intense heat.[8]

Essentially all of the nation's hard coal came from four large veins in the mountains around Scranton and Wilkes-Bare in eastern Pennsylvania. Anthracite's clean, slow burn helped make it popular among domestic users, as did its hard exterior, which made it less dirty to handle before burning. Hard coal could burn long hours without tending, as well, meaning that residents could leave a fire overnight in a stove, fireplace, or furnace and find it still hot in the morning. But even in the large cities nearest the anthracite mines, Philadelphia and New York, hard coal commanded a significantly higher price than did the more abundant bituminous. And, in more distant ports, such as St. Louis, anthracite could cost more than four times as much per ton as bituminous from nearby Illinois mines.

Soft coal's wide distribution gave it a competitive advantage over anthracite. By 1900 miners dug bituminous in over twenty states, meaning that most major cities had easy access to inexpensive soft coal. However, it compared unfavorably for domestic use. It was dusty to store and smoky and smelly when burned. Bituminous fires required closer attention, because they left more ash and hard, unburned residues of iron and other impurities called clinkers. Users needed considerable expertise to prepare a bituminous fire that would burn all night without further tending. Still, despite its shortcomings as a domestic fuel, bituminous greatly outsold anthracite in the nation after 1870. Soft coal, due to its availability and low cost, became the fuel of choice for most industry west of the Appalachian mountains, and the nation's railroads, major consumers of coal, also turned to bituminous fuel as wood became prohibitively expensive. The relative scarcity of "smokeless" anthracite ensured that the continuing industrialization of America would be fueled largely by its dirtier relation, bituminous.[9]

Aside from choosing between soft and hard coals, consumers also selected from a wide variety of fuel sizes, each of which carried certain advantages and disadvantages, and, of course, its own price. Consumers could purchase coal in large blocks, in lumps, or in the smaller stove, egg, chestnut, pea, and buckwheat sizes. The type of furnace or stove in which the coal would be burned largely determined which size the consumer bought, for even though pea coal cost less than lump, if it fell through the grating of the furnace, it did the purchaser little good. Of course the size did not necessarily speak to the quality of the coal. Some mines could produce up to fifteen different sizes, all processed and sorted at the mine's mouth. Thus customers also knew the names of the mines which produced the highest quality coals, those that burned hottest and longest. St. Louis consumers, for example, knew Big Muddy coal was the best Illinois had to offer, and they paid extra for it. Similarly, industry knew Connellsville coke bested all others on the market, and that reputation helped sell the Pennsylvania fuel all around the nation, despite its relatively high price.[10]

Given the high costs of shipping both coal and coke, industries that required large amounts of energy tended to locate near productive mines. The most important coal field, covering much of western Pennsylvania, West Virginia, eastern Ohio, and eastern Kentucky, supplied Pittsburgh, Cleveland, Cincinnati, and many other, smaller industrial cities with inexpensive bituminous coal. Another important field, running from central Illinois down to western Indiana and western Kentucky, supplied bituminous to nearby Chicago, St. Louis, Louisville, and Indianapolis. And by 1900 a significant field in northern Alabama and central Tennessee had helped turn Birmingham into a thriving steel city and spurred the growth of Tennessee's industry in Chattanooga, Nashville, and Memphis. Not all coal found its way into local or even regional markets, however. Pennsylvania's anthracite coal, for example, entered markets all over the nation, moving northeast to Boston, south to Baltimore, and west to Chicago. In fact, exports of both anthracite and bituminous helped make Pennsylvania the largest producer of coal in the nation. In 1910, Pennsylvania mined 47 percent of the nation's coal, nearly two-thirds of it bituminous.[11]

Although the coal trade moved fuel around the nation by rail, ship, and barge, the high cost of transporting coal ensured the development of industrial cities in close proximity to the great fields.[12] Many of these cities became remarkably fuel intensive, as inexpensive coal attracted industries which consumed great quantities of fuel. Pittsburgh, for example, located in one of the world's most energy-rich environments, consumed 15.6 million tons of coal in 1910, or more than 29 tons per person. That year the city of Pittsburgh consumed almost as much coal as did New York City, which was nearly ten times as populous. The low cost of coal at the mine mouth ensured high local demand, and intense energy use in cities close to bituminous fields greatly exacerbated their smoke problems.[13]

In the Midwest the cost differential between local bituminous and imported smokeless coals secured success for even terribly dirty fuel. The St. Louis coal market, for example, offered a wide variety of coals: Pennsylvania anthracite, West Virginia's high-quality Pocahontas bituminous, Illinois's Big Muddy bituminous, and the smoky, low-grade Mt. Olive coal from just across the Mississippi. Even though Pocahontas coal burned efficiently and almost smokelessly, it cost more than twice as much to use in the boiler room. For a factory consuming hundreds of tons of coal per week, even small differences in price between two types of coal could hold considerable meaning. For railroads, burning thousands of tons per week, finding the least expensive coal could mean the difference between profit and loss. Thus, in St.

Louis and other midwestern cities the smoke of dirty local coals fouled the air, despite the availability of cleaner alternatives.[14]

Although cleaner coal alternatives reached the markets of industrial cities, many midwestern cities had no significant non-coal alternatives well into the twentieth century. With nineteenth-century oil discoveries in Pennsylvania and Ohio various petroleum products began to make important contributions to the nation's fuel market, but in 1910 oil represented only 6.1 percent of the nation's energy consumption, as measured in Btu's. Before World War I, petroleum supplied kerosine for lamp lighting, gasoline for automobiles, and fuel oil for some railroads and industry. Neither gasoline nor kerosine competed with coal, and fuel oil became an important energy source only in regions near western oil fields, particularly in California and Texas, where distance from coal fields made the liquid fuel economical. Not until after World War I would fuel oil begin to seriously compete with coal in the larger eastern and midwestern fuel markets.[15]

Natural gas also eventually competed with coal, but into the 1920s shipment and storage problems greatly limited the use of the clean-burning fuel. The discovery of a large store of natural gas in Murrysville, Pennsylvania in 1883 led to a brief spike in the use of the fuel in Pittsburgh, but for the most part natural gas played only a small role before World War I, and it represented only 3.3 percent of the energy market in 1910. This small share placed natural gas on a par with hydropower, which also continued to have a very limited impact on urban energy markets.[16]

Perhaps the limited nature of non-coal energy alternatives is best revealed in the continuing importance of fuel wood, which in 1910 represented the second largest source of Btu's in the nation, providing 10.7 percent of the energy market. Once an important energy source for homes, factories, foundries, and railroads, by the 1910s wood had become a largely rural fuel, still supplying heat and cooking fuel for homes outside the largest cities. For the most part, however, relatively high prices prevented wood from competing with coal in urban areas. As the United States continued to urbanize, fuel wood continued to lose its relative importance, representing just over 5 percent of the energy market by 1940.[17]

Although coal consumption surged in urban America in the late 1800s and through the 1900s, its use predated the nation's rapid industrialization after the Civil War. As early as 1830 Pittsburgh relied on coal for both domestic and industrial purposes. By the 1840s, anthracite had entered the markets of Philadelphia, New York, and other industrial towns, fueling a rapid expansion of manufacturing. As the new coal supplies offered cities inexpensive fuel, energy-intensive industries such as iron manufacturing moved into cities and away from their former energy supplies, forests. As the historian Alfred Chandler has argued, coal, specifically the anthracite coal that made its way into eastern markets, helped spark the industrial revolution in the United States by providing a higher quality fuel for iron and steel manufacturers. Although wood and water power played important roles in fueling early American industries, the creation of great industrial cities awaited the availability of a more abundant and flexible fuel. Essentially the growth of manufactures in American cities had been curtailed by the limited nature of their fuel supplies. As coal gradually replaced wood, not just in heavy industry but in all sectors of the economy, it fueled not only industrialization but also urbanization. The new availability of coal energy removed one of the most important barriers to the expansion of cities. As the well-known engineer William Goss said of industrial Chicago, "It is not that coal is burned in this locality or that; it is burned everywhere in the city. The city's existence is predicated on the consumption of fuel."[18]

While coal provided crucial energy to young industries in antebellum cities, the rapid expansion of coal consumption occurred only after the Civil War. In what historians often call the coal age, the nation's consumption expanded from just over 20 million tons in 1860 to over 650 million tons in the peak year of 1918. In other words, in sixty years the nation's coal consumption grew by more than 3,200 percent. Consumption expanded so fast that in 1908 the United States Geological Survey declared that the consumption of the preceding decade had exceeded that of the entire previous century. Although the use of other fuels, particularly petroleum and natural gas, also expanded rapidly during these decades, in the 1910s coal supplied more than 75 percent of the nation's total energy.[19]

Just as important, the anthracite supply could not keep pace with bituminous production. So, whereas the relatively smokeless anthracite constituted 54 percent of the coal consumed in 1860, it made up only 17 percent of the coal burned in 1918. Clearly, the limited supply of smokeless coal and its geographical concentration in eastern Pennsylvania ensured that industrialization in the Midwest and the South would be fueled by other energy sources. By the turn of the century, even the eastern markets relied heavily on the dirty bituminous coal of the Appalachian fields. "One by one the great cities of the East are being devoured by the black smoke beast," wrote the *Medical Record* in 1906; "even tidy Philadelphia is becoming grimy and soot-soiled." As early as 1899, Bostonians began to speak out against the introduction of soft coal and the smoke that came with it. Through the decades be-

TABLE 1
The Making of King Coal:
Apparent Consumption

Year	Bituminous coal	Anthracite coal
1870	20,817,000	19,822,000
1880	51,036,000	28,210,000
1890	110,785,000	45,614,000
1900	207,275,000	55,515,000
1910	406,633,000	81,110,000
1920	508,595,000	85,786,000

NOTE: Rounded to thousands of net tons
SOURCE: Adapted from Sam H. Schurr and Bruce C. Netschert, *Energy in the American Economy, 1850–1975* (Baltimore: Johns Hopkins Press, 1960), 508–9.

fore World War I, as America's industrial cities consumed more and more coal, they also consumed more and more of the smokiest varieties.[20] (See table 1.)

While the expansion of coal consumption in the United States was extraordinary, American production relative to the rest of the world was just as impressive. In 1913, the Geological Survey estimated the entire world's coal output at under 1.45 billion tons. Of that amount the United States contributed .57 billion tons, or nearly 40 percent. Great Britain was the next largest producer with .32 billion tons, or approximately 22 percent. Just fifteen years earlier Great Britain had outproduced the United States. Perhaps no other set of statistics, not even those concerning steel and iron production, so dramatically reveal America's surge to leadership in the world economy. As one author noted in 1897 (when Great Britain still outproduced American mines), "The most polished nations cannot for the future dispense with coal, and the degree of a country's civilization may almost be estimated by the quantity of this combustible which it consumes."[21]

The consequence of coal for industrializing America attracted considerable comment. Urban residents of all classes well understood the significance of the black fuel. Local businessmen and politicians in coal-dependent cities praised their proximity to the mines that fed their markets and their factories. Even amidst a diatribe against smoke, Henry Obermeyer paid tribute to King Coal in 1933, noting that the fuel was the foundation of the nation's wealth. "Almost the sum total of modern construction in American is nothing more than a monument to coal," he announced, lamenting the lack of less polluting alternatives.[22] Many of coal's most enthusiastic proponents noted the special relationship it had not only with industrialization around the world but also with the development of a new epoch. A curator of mineral technology at the Smithsonian Museum, Chester Gilbert, summarized the role of coal in American society: "Coal in short is the nucleus around which the material growth of civilization builds." According to Gilbert, coal was fundamental. Modern civilization arose from the "organized employment of mechanical energy," and at the turn of the century coal supplied that energy. Another commentator, Chautauqua lecturer Charles Barnard, concluded, "It may safely be said that our whole civilization now rests on steam-power, and the comfort and safety of the nation depend on the amount of coal we have in our hills and mountains." Gifford Pinchot, the renowned conservationist crusader, also exalted coal as he lobbied for more efficient mining and burning. Although the American supply of coal was large, Pinchot emphasized that it was finite. More important, according to Pinchot coal was "in a sense the vital essence of our civilization." Like most Americans, Pinchot could hardly envision a future without coal.[23]

At least one observer, Robert Bruere, placed coal at the center not only of American civilization but of a world civilization to come. In 1922 Bruere wrote, "With the coming of coal and coal-driven machinery the earth and the fullness thereof was unlocked for the service of man." Americans had used coal to create surpluses in wealth heretofore unseen in world history. From an optimistic, Christian viewpoint Bruere saw glorious potential on earth: "There was not only the possibility of the good life for each but also of a noble, well-ordered civilization for all." Coal, he thought, embodied "the chance of a world civilization," where the elimination of scarcity would leave only peace and cooperation among the nations of the world.[24]

Both the psychological attachment to coal and its economic importance created significant barriers for those who hoped to mitigate the most troubling aspect of its use: smoke. Since most coal created smoke when burned, many Americans had come to accept smoke as a necessary part of coal consumption. Thus, through the nineteenth century Americans had come to associate smoke with all those positive changes attributed to coal: production, prosperity, and progress. For many residents of cities reliant on dirty soft coal, smoke streams rose from factory stacks, from locomotives, and steamships, like banners of civilization. In an age of economic booms and busts, all too often workers saw idle smokestacks as evil omens, indications that no work could be had. Some city boosters went so far as to claim that smoke helped attract workers and businesses to their cities. "I came to Birmingham in 1886 because there was smoke here," the Birmingham *Age-Herald* quoted one Alabama laborer as saying in 1913. He insisted that after the depression of the 1890s, "the smoke coming from the big mills sent a thrill of encouragement and optimism through every man in Birmingham." If some workers could thrill at the sight of smoke, others could simply accept it as part of urban life. "The majority of the people are wage earners who don't object to smoke, as they have to put up with it in making a livelihood," said another Birmingham wage earner. Of course workers in the steel factories around Birmingham were not the only Americans to accept the smoke of their cities.[25]

Nor were laborers alone in their appreciation of smoke. In Chicago, coal dealer William Rend declared smoke beneficial to the city. "The Creator who made coal knew that there would be smoke," Rend announced, "and knew that smoke would be a good thing for the world." Although Rend was unusually enthusiastic for coal, a businessman did not need to have immediate interest in the coal market to find some glory in smoke. In his 1914 novel *The Turmoil*, Booth Tarkington summarized nicely the philosophy behind smoke's positive image. "It's good! It's good," shouted the novel's wealthy businessman, Sheridan. "Good, clean soot: it's our life-blood, God bless it!," he laughed to women who had hoped he would help abate the city's smoke. "Smoke's what brings your husbands' money home on Saturday night. . . . You go home and ask your husbands what smoke puts in their pockets out o' the pay-roll—and you'll come around next time to get me to turn out more smoke instead o' chokin' it off!" Like the fictional Sheridan, workers, managers, and owners in coal-dependent cities—particularly those dependent on bituminous coal, including Chicago, Pittsburgh, St. Louis, Cincinnati, and Birmingham—could look upon their smoky atmospheres with acceptance, if not pleasure, well into the twentieth century.[26]

In this regard, turn-of-the-century attitudes toward smoke differed significantly from those concerning other forms of pollution. William Barr, a Cleveland engineer and a student of the smoke problem, commented in 1882, "There are some things which must be endured because they cannot be cured, and I have no doubt that many persons look upon smoke as a sort of necessary evil; others rather take pride in it, and imagine smoke to be an index to local activity and enterprise." The first task of those interested in controlling smoke, then, was to change the public's perception of it, to convince fellow urbanites that the smoke with which they lived was neither necessary nor an indicator of progress. The second task would be to convince a society heavily dependent on coal that an attack on smoke was not an attack on coal, that smoke abatement did not require an abandonment of the energy source that had fueled the creation of the American industrial civilization.[27]

Progressive urbanites, even those most actively engaged in the fight against smoke, were not likely to abandon industry, urban life, or the coal that fed them both. The perfected civilization they envisioned was populated by prosperous cities, abundant surpluses, certain morality, beauty and health. Unfortunately, in the late 1800s, the industrial order that had brought prosperity and surplus had also jeopardized the moral order and compromised beauty and health. The smoke palls that shrouded many of the nation's prominent cities reflected the connection between progress and ugliness, and symbolized the unhealthfulness and impurity of the new industrial city. The primary goal of antismoke activists, then, was to convince coal-reliant Americans that smoke was uncivilized and unprogressive, just the opposite of what many had come to believe. It would not be an easy task.

The difficulty of the task became clear in New York during the long coal strike of 1902. Facing shortages of anthracite coal, New Yorkers turned to soft coal as never before. The dramatic increase in the burning of bituminous caused an equally dramatic increase in smoke, and complaints flooded the city's board of health. New Yorkers were not accustomed to the thick palls of smoke that began to haunt the city, and by mid-June the New York *Times* announced that the smoke nuisance had reached a crisis. A New York *Tribune* headline read, "A Cloud Over the City," signaling not just the physical condition of the metropolis, but the psychological condition of its citizens as well.[28]

In less than a month's time, the anthracite strike wreaked havoc on New York's environment. The weather bureau claimed New York's air was as impure as it had ever been, and the heavy smoke began to cause a wide variety of problems. The sales of Panama hats plummeted as would-be purchasers noticed how poorly the light straw fared in New York's newly sooty atmosphere. One laundry reported a one-third increase in business due to the rapid soiling of clothing. Doctors and laymen expressed fears concerning the effects of the smoke on eyes and lungs and on human health in general. By June 13, low visibility due to smoke menaced shipping in New York Harbor as "ferryboats, tugboats, steamboats—every class of craft—were belching clouds of soot from their stacks," and the *Times* reported that a "bank of black smoke" lay over the water.[29] On June 14, a local union of butchers complained about the smoke from the elevated railroad on Manhattan. Chairman William Wollman of the Amalgamated Meat Cutters said, "The smoke rolls down in dense clouds and not only spoils the looks of the meat, but makes a distinct difference in its taste, certainly not for the better." R. C. Thomson, visiting New York from the nation's smokiest city, Pittsburgh, sat in a cafe and watched the soot settling on the white table cloth, onto the butter and into his cream. "You don't like it, I see," he said to a reporter. "But you might as well get used to it. Shut your eyes when you pour your cream."[30]

New Yorkers were not about to shut their eyes to the smoke crisis, however. Residents had long expressed great pride in the purity of their air, and many New Yorkers saw in the smoke a threat to the beauty of the city, the health of its citizens, and the morality of the civilization that united the two. New York City had a longstanding and rigorously enforced antismoke ordinance. Although it only prohibited the production of dense smoke from the burning of soft coal, in practice its enforcement had prevented most coal users from burning soft coal at all. Limited by available technology, New Yorkers found it nearly impossible to burn soft coal without creating a thick, black smoke. As the strike began, then, the city remained heavily dependent on the slightly more expensive, but much cleaner-burning anthracite. But as the price of anthracite rose sharply during the strike, owners of hotels, office buildings, apartment buildings, factories, and tugboats, and the managers of commuter railroads, pumping stations, and electric power stations, all turned to soft coal and risked arrest.[31]

With calls to action coming from all quarters of the city, arrests came quickly. Just five days after the strike began, a delegate of the Blue Stone Cutters submitted a resolution at a Central Federated Union meeting which called upon the city "to enforce the soft-coal ordinance to protect public health and to arouse the public sentiment against the operators." The delegate explained further, "If soft coal is to be used generally here it will increase the death rate and spoil our beautiful city." Although motivated by union solidarity, the resolution of the Central Federated Union echoed the arguments of middle-class New Yorkers who thought smoke threatened their health and knew it spoiled the aesthetics of their city. Residents feared New York would join the ranks of the nation's dirty cities and lose its reputation as a fine residential metropolis. Andrew Carnegie told reporters outside his Fifth Avenue mansion, "If New York allows bituminous coal to get a foothold, the city will lose one of her most important claims to pre-eminence among the world's great cities, her pure atmosphere."[32]

To avoid such a fate, Dr. Ernest Lederle, commissioner of New York's Board of Health, began making arrests for smoke violations just two weeks after the strike began. Only two weeks later the health department had issued 150 complaints against soft coal users, and the police had arrested 25 offenders. Those arrested faced the prospect of fines from $50 to $250 if convicted. By late June, the courts began to clear the smoke cases. In sentencing the Riverside Cold Storage Company to a fine of $50, a city judge said, "The condition of our streets is horrible. Owing to the smoke nuisance the soot descends on furniture, books, and other household articles, damaging and discoloring them, and the conditions are rapidly becoming intolerable." Defendants could only claim that they could no longer find or afford anthracite. Prosecutors only needed to produce one coal retailer willing to announce he had anthracite for sale, at a fair price, to prove their cases.[33]

City officials suspected that many of those who switched to soft coal did so not because of a real shortage of anthracite but because they could use the strike as an excuse to purchase less expensive bituminous coal and save money on their fuel bills. Yet for most consumers the shortage of anthracite was real enough. The railroad companies that controlled the anthracite mines ceased delivery to coal dealers as soon as the strike began, in an effort to save coal for their own locomotives or to await higher prices. Retailers also withheld anthracite until prices surged. Each new rumor of an impending strike settlement brought more coal onto the market, as retailers hoped to fetch the highest prices. Each new realization of the continuation of the strike forced New York's largest consumers to search farther and wider for new sources of clean-burning coal. By July 25, anthracite garnered $8 per ton; a month later it reached $9.50. By late September, anthracite topped $16 per ton, more than three times its prestrike price. Some residents turned to wood as an alternative for home fires; some businesses turned to relatively clean British coal, from the mines of Wales and Scotland, as a stopgap measure.[34]

As the strike reached into its sixth month, The Nation declared that the mining interests had put the East in a state of siege. "The march of the seasons has not been stayed. Winter is upon us, and our Eastern population is preparing to practise those economies and face those hardships which were the lot of the first settlers. This means a turning back of civilization." As the strike persisted and the city starved for fuel, schools failed to open and hospitals struggled to keep warm. The extended strike, The Nation remarked, had pauperized the entire city. In mid-October, New York's mayor suspended enforcement of the smoke ordinance as the scarcity of anthracite made the burning of bituminous coal a necessity. New Yorkers and New York's air had finally succumbed to the strike.[35]

Many New Yorkers lamented the loss of their pure air and cursed the spoiling soot. But, after all, what could be done? The city required fuel. If the city could find no clean fuel, then it must burn the dirty coal. The city appeared to have little choice. No one had the power to clear the skies. The industrial city relied on complex systems beyond the control of any one group.

Failures in one system a hundred miles away, in the hills of Pennsylvania, had real and immediate consequences for the residents of New York. The growing smoke cloud revealed a growing inability of city residents to control their environment. New York could find no solution to its smoke problem and joined the list of other industrial cities, both American and European, that had sacrificed the purity of their air in the pursuit of prosperity. The city hunkered under its black shroud until federal involvement helped resolve the strike and anthracite returned to city markets.[36]

In the end, New Yorkers discovered that they had owed their pure industrial air not so much to adequate laws and enforcement, but to the city's proximity to Pennsylvania's large stores of anthracite coal. The shortage of anthracite had made a mockery of the law and the ability of the municipal government to regulate the city's environment. The smoke also gave ample evidence of the centrality of fuel type in determining air quality. As historian Peter Brimblecombe has written concerning medieval Europe, "a history of air pollution is almost a history of fuel." Still, some New Yorkers did hold out hope that the city might regain its pure air, even if the use of soft coal continued. As the city struggled to keep bituminous consumption low, officials frequently announced that soft coal could be burned without smoke, borrowing the rhetoric used by antismoke activists in cities where bituminous dominated the fuel markets. Quoting "competent engineers," newspapers tried to educate readers on how to use soft coal without producing offensive smoke. If proper handling of the fire did not prevent the smoke entirely, the papers noted, several "devices" could "consume" the remaining smoke.[37] Many residents of Manhattan also lobbied for a long-term solution to the smoke from the elevated railroads: electrification. Although elevated roads had begun electrification before the strike began, smoke from elevated locomotives, particularly offensive given their proximity to upper-story windows of residences and offices, had solidified public support for a rapid transition to clean electric power.[38]

Most New Yorkers continued to exhibit a faith in the ability of technology to improve their lives, even while understanding that many of the technological advances that had so enriched their lives had also compromised their environment. Industrial expansion had improved incomes and living standards and expanded the middle class, but had also fouled the city's air and waterways. Modern transportation systems had allowed the city to expand, but had also added considerable smoke and noise and a degree of danger to the streets. Most progressive reformers believed that only further technological advances could cure the environmental problems that the use of technology had caused. Or, in the terminology of the day, the only "cure for the evils of civilization is more civilization."[39]

By the late fall of 1902, the end of the strike brought a gradual end to the smoke crisis, as anthracite prices dropped and consumption of soft coal eased. But New York's acute smoke problem simply became chronic, as many large consumers of coal continued to use soft coal and reap the economic windfall that came from using the cheaper fuel. Although New York's smoke problem never approached those of the Midwest's industrial centers, its citizens' reactions to the "smoke evil" resembled those of their inland counterparts. Activists created antismoke organizations, sponsored scientific studies, supported smoke abatement legislation, and lobbied for its enforcement in a crusade for civilized air.[40] New Yorkers joined Chicagoans, Cincinnatians, Pittsburghers, and others, in praising their good economic fortunes, cursing the poor state of their environment, and reforming the half-built civilization that had created both. If a vision of a higher civilization remained obscured, the dense smoke clouds of the city deserved considerable blame.

two

Hell Is a City
Living with Smoke

Hell is a city much like London—
  a populous and a smoky city;
There are all sorts of people un-done,
And there is little or no fun done;
Small justice shown, and still less pity.
     —Percy Bysshe Shelley, from "Peter Bell the Third" (1819)

Smoke—The visible exhalation, vapor, or substance that es-
capes, or is expelled, from a burning body, especially from burning vegetable matter, as wood, coal, peat, or the like.
     —*Webster's Dictionary*, 1895

As the United States rapidly developed into the world's largest economy in the decades following the Civil War, American coal fueled the creation of a new civilization. By the turn of the century, the United States outpaced Great Britain in both coal and steel production. Inland industrial cities, such as Chicago, Pittsburgh, Cleveland, and St. Louis, experienced exponential growth in population, production, and consumption. These and other cities entered a new era of prosperity and expansion. But even as the nation created products and profits at unprecedented levels, much about American cities suggested chaos. In the city the contradictions of industrial society lay bare. The nation's urban economies boomed and yet poverty abounded. New prosperity brought skyscrapers, electricity, improved transportation systems, and countless other advances to urban living, but at the same time tenement districts bred disease and discontent, and the urban environment suffered under the weight of rapid population growth and polluting industries. Even as the nation approached world preeminence in production and commerce, American cities still struggled to collect garbage, remove sewage, supply potable water, and protect clean air. In the wealthiest nation on earth, the cities looked impoverished and chaotic.[1]

Smoke symbolized the contradictions of the city. To many urban residents smoke meant progress and jobs. At the same time, smoke was filthy and oppressive. To many visitors smoke was central to the industrial city scene, obvious upon approach, dominating the sky and atmosphere. As the American author Waldo Frank wrote of Chicago in 1919, "The sky is a stain: the air is streaked with runnings of grease and smoke. Blanketing the prairie, this fall of filth, like black snow—a storm that does not stop. . . . Chimneys stand over the world, and belch blackness upon it. There is no sky now." Here, in the dense and shifting clouds of carbon and sulfur, drifted visible evidence of the contradictions of progress, urban life, and the new civilization. And while the black smoke obscured the present, as urban residents squinted with stinging eyes, the smoke could also obscure the future. To what end would this dirty civilization progress?[2]

Of course smoke could never be just a symbol of the contradictions inherent in industrial urbanization; it was too real, too palpable. As one eloquent Milwaukeean wrote of smoke in 1888: "It penetrates our houses, it befouls the atmosphere, spoils everything, benefits nothing. . . . My clothes are dirtied by this smoke. I swallow it. It fills my eyes, chokes my bronchial tubes. It comes between me and the sun and I see my fellow beings suffer day by day." Surely this was no mean nuisance. On still, heavy days choking smoke clouds gathered in the nation's cities, suggesting a coming storm that never arrived.[3]

Perhaps no voice spoke out against the smoke problem more consistently and articulately than the New York *Tribune* in the late 1800s. In the relatively clean air of New York City, the black smoke of soft coal appeared as a foul intruder. One editor judged permitting a smokestack to emit dark smoke "the blackest of infamies." In the winter of 1898, an editorial titled "The Doom of Darkness" continued the paper's campaign against the increasing smoke in the city. It recounted a morning walk on a crisp, clear winter day, under a "brilliant azure" sky and the energizing "pureness and exhilarating buoyancy of the air." But this enjoyment was cut short. "There was a long black cloud incessantly streaming eastward on the wind, like a river of darkness flowing over the city. Hour after hour it flowed on, unbroken." The smoke stream persisted, "in violation not only of the laws of Nature, but of the laws of man," and gave ominous foreshadowing for the years to come. In the spring of the next year, the editor lamented, "Day by day the air becomes blacker and fouler and more suffocating." The author made clear his understanding of what exactly was happening in New York: "[H]ere in this city, for which nature has done more than for most other great cities of the world, men are deliberately defiling earth and air and sky with blackness."[4]

Unlike New York, where smoke was an unwelcome intruder, in many cities smoke was a full-fledged citizen, native born and always at home. The volume of smoke produced in some of these cities was astonishing. Since researchers had little ability to measure the amount of smoke itself, many turned to the study of soot as a representation of the density of the smoke above. In 1912, as part of a larger study of the smoke problem in Pittsburgh, the Mellon Institute measured soot fall in various locations in the city. The study concluded that in some areas nearly 2,000 tons fell per square mile annually. In total for 1912, 42,683 tons of soot blanketed the city. At an exhibit organized by Pittsburgh's Smoke and Dust Abatement League, the Mellon Institute displayed a replica of the Washington Monument with a similarly shaped obelisk representing the total soot fall in Pittsburgh. In this very graphic display, the black tower of soot loomed over its smaller white rival.[5]

Studies of soot fall also confirmed that not all city residents suffered equally under the smoke clouds. In Cincinnati, researchers found that soot fell most heavily in the central business district, where in 1916 an estimated 217 tons fell per square mile. Meanwhile, in Cincinnati's close but elevated suburbs, the soot fall did not reach twenty tons per square mile. Residents who lived in the tenement districts around downtown and in the working-class neighborhoods near industries bore the worst of the smoke nuisance. But, while smoke and soot were worst in central cities, near railroad stations, steamship docks, and factories, those who complained most bitterly often lived beyond the palls and looked upon the clouds from some short distance. A middle-class resident of Cincinnati's Clifton neighborhood, where soot fall was only a small fraction of that of the central business district, might feel minimal effects from the smoke at home, but still have a good understanding of the magnitude of the city's smoke problem. From a perspective outside the dense smoke pall, the gravity of the problem was clear. Often from suburban heights, middle-class urbanites peered through the smoke haze upon their productive creations and longed to express greater pride.[6]

Taking lessons from more recent air pollution problems, some historians have noted that smoke became a nuisance where temperature inversions were common. Inversions occur when incoming warmer air traps cooler air below, and they are particularly common and persistent in hilly regions, where cool air can be trapped in valleys, such as in Pittsburgh or Cincinnati. A "ceiling" created by an inversion can impede the course of warm, rising smoke. In this situation smoke could build in the city's atmosphere, creating

1. View N. West from Roof of Union Station from a Point on North Side of Building
. . . (20 June 1906). Pittsburgh flexes its industrial muscle. Courtesy of the Carnegie
Library, Pittsburgh.

heavy palls. During one such inversion in Pittsburgh, a visitor looking down
from a nearby hill described the city as "hell with the lid taken off," as he
peered through a heavy, shifting blanket of smoke that hid everything but
the bare flames of the coke furnaces that surrounded the town. Still, while
weather did greatly affect its density and distribution, smoke could become
a nuisance even without the trapping effect of a temperature inversion. The
relatively heavy particles in coal smoke did not need special weather conditions to impede their dispersal, particularly in the late 1800s, when chimneys
and smokestacks barely exceeded the height of the structures they served.
Except in windy conditions, dense smoke hovered near its producers and
soot fell heaviest where production was highest.[7]

While the sheer quantity of smoke posed some problems, particularly regarding visibility, qualities other than its opaqueness caused even more se-

rious problems. Coal soot, for example, was particularly invidious, for it not only coated everything in the city with black dust, it also had an oily quality, which helped it cling to clothing, curtains, furniture, and other items. It smeared and stained. Soot could stick to exposed skin, collect in nostrils, lungs, eyes, and stomachs. It clung to buildings, walls, books, and dishes, and could not be simply brushed away. Soot found its way into cupboards and closets, attics and cellars, and it colored the cheeks of the city's children as they played in the dusty streets.[8]

The density of smoke caused significant health problems. Several lung ailments consistently led among causes of death in the nation's industrial cities, and no doubt the foul air contributed to the death rate. Tuberculosis headed the list of urban killers in the decades surrounding the turn of the century, but although smoke certainly made the lives of consumptives less comfortable, and perhaps shorter, the death rate from tuberculosis had little if anything to do with coal smoke. Rather, housing conditions, particularly the density and filthiness of tenement districts, were much more significant factors in the spread of the tubercle bacillus. Death rates from pneumonia, bronchitis, and asthma, on the other hand, were affected by smoke. Although these three diseases are rarely fatal today, in the late 1800s all three were serious killers. In Cincinnati, for example, the three leading causes of death in 1886 were tuberculosis, pneumonia, and bronchitis. In total, 31 percent of all deaths in Cincinnati that year were lung related. The true extent of smoke's effect on health in these decades remained unknown, but the blackened lungs of urban cadavers indicated the depth of the problem.[9]

Although the aesthetic aspect of the smoke problem may appear insignificant compared with the health effects, for residents of turn-of-the-century cities, the visual impact of smoke was much more obvious and immediate than the threats to health. Particularly in the first decades of the twentieth century, when progressives worked toward a general beautification of their cities, controlling smoke became extremely important. Led by reformers, such as Charles Mulford Robinson and J. Horace McFarland, the City Beautiful movement of the early twentieth century attempted to use city planning to beautify and civilize American cities. City Beautiful plans typically contained neoclassical architecture for grouped public buildings, grand public spaces, particularly formal parks, improved street designs, including parkways and street trees, and the control of the many urban pollutants, including smoke. As the Chicago *Record-Herald* noted in 1911, "A filthy city cannot be beautiful. Smoke, soot and cinders render every attempt at adornment a hollow mockery." In Chicago smoke held special meaning for City Beautiful endeavors. The Illinois Central, one of the city's busiest lines, ran through Grant Park along the city's shoreline to a downtown terminal. For decades the smoke and cinders from Illinois Central locomotives kept visitors out of the park and even clouded out views of the lake from adjacent property. Still, not just Chicago but all coal-reliant cities faced aesthetic challenges. As the Cleveland Chamber of Commerce concluded, "The presence of coal smoke in large quantities constitutes perhaps the greatest hindrance to the highest development of civic beauty and refinement."[10]

Soot also had aesthetic effects. City residents often complained of soiled buildings. The new, large, white, neoclassical structures, so important to the City Beautiful conception of architectural maturity, suffered most severely. Chicago's creation of the "White City" for the 1894 World's Fair indicated the highest in civic achievement. A clean, pure, well-planned, impressive, and white city set the standard at the turn of the century. Chicago's White City, constructed largely from flammable materials painted to resemble stone, did not survive a series of fires over the next few years, and attempts at its replication in other cities, through the construction of impressive, neoclassical municipal buildings, did not survive years of soot. Public buildings, courthouses, libraries, capitals, all gathered soot on their stone and marble walls. Their darkening hues suggested early decay, just as surely as the gradual destruction of Chicago's White City symbolized the impossibility of lasting purity in an impure environment.[11]

Smoke darkened not just buildings, but the entire sky. Smoke clouds cast a somber shadow over industrial cities. The psychologist J. E. Wallace Wallin contended that smoke displaced "the cathedrals of nature with an uninteresting, nasty, black opaque pall of soot which stimulates tendencies toward discontent and frequently arouses morbid emotions." In Chicago, the dark smoke certainly aroused discontent among artists, who complained that it had "polluted the pure atmosphere so necessary for the artistic temperament." Actually, many observers suggested that diminished sunlight had more serious implications, as physicians noted that the ultraviolet rays in sunshine killed bacteria and were generally important for a healthful environment. For this reason some physicians listed smoke as an indirect cause of higher death rates in smoke-shrouded, disease-ridden tenements.[12]

Smoke could also affect the weather itself, not just by darkening the skies, but by changing the chemistry of the atmosphere. Perhaps no city was more famous for its altered weather than London, where obstinate fogs could engulf the city for weeks at a time. Of course London was prone to hazy weather, but coal smoke made the fogs thicker and more persistent, and, not trivially, frequently deadly. Pittsburgh gained a similar reputation for

2. Soot stained the exteriors of all types of buildings. Here, later in the century, workers clean the exterior of the Oliver Building in Pittsburgh. Courtesy of the Carnegie Library, Pittsburgh.

smoke-induced gloomy weather. A meteorologist at the Mellon Institute concluded that smoke made city fogs more persistent than country fogs, decreasing the intensity and length of sunshine in cities. Smoke may also have acted as a blanket, trapping heat and keeping cities warmer than the surrounding countryside.[13]

At least one researcher, Swedish scientist Svante Arrhenius, suggested even more serious implications of gathering carbon dioxide clouds spewed by coal fires. In 1896, while investigating long-term temperature changes on earth, Arrhenius concluded that massive changes in the carbon dioxide content of the atmosphere could cause significant increases in surface temperatures. Charles Van Hise, the well-known University of Wisconsin scholar, repeated Arrhenius's argument in his popular work on conservation, giving the idea of global warming due to human combustion of fossil fuels a broad American audience in 1910. By 1912, *Scientific American* suggested that recent uncommonly warm summers may have been due to the continuing exponential growth in coal consumption, repeating Van Hise's warning that human activity might have climatological consequences.[14]

Of course smoke had much more obvious and immediate effects on the natural world, particularly on vegetation in and around cities. In 1906, Andrew Meyer, the city forester of St. Louis, estimated that three-fourths of the trees lost in the city the previous year had died from smoke-related problems. Meyer particularly feared the continuing loss of old hardwoods in the city's Forest Park. The accumulation of soot on leaves and the presence of sulfurous gasses proved highly toxic for some plants. Urbanites frequently lamented the loss of conifers within their cities. While some plants could endure the smoky weather, others fared poorly, and the loss of diversity, particularly among flowering plants, caused many urban residents to curse the smoke. As one woman living in the shadow of Cleveland's American Steel and Wire Company complained in 1905, "At night we can hear the cinders falling like hail on the roof. Nothing will grow here. My trees and flowers are dead." Cities could hardly expect to improve their landscapes with flowers and trees if neither could long survive the thick atmosphere.[15]

In the late 1800s, middle-class urbanites placed great value on health and aesthetics not only for their direct influences on urban life, but also because they affected morality. By the 1890s, reformers believed that the urban environment had a great influence over the character of its residents, and many outspoken critics of smoke connected the dirty environment with the deterioration of morality. In responding to a study of the psychological aspects of the smoke problem in Pittsburgh, Dr. Max Witte concluded, "I have no doubt that the murky, smoky atmosphere of your city exerts an influence to a greater or lesser degree upon the morals and disposition of the young people who dwell habitually in your climate." One of Cincinnati's prominent physicians, Dr. Charles Reed, commented, "Physical dirt is close akin to moral dirt and both combined lead to degeneracy." Many commentators noted the close relationship between cleanliness and Godliness, and middle-class urbanites often articulated a concern for the moral influence of filth on the city's poor, particularly poor children. In an effort to arouse public sentiment, the superintendent of the Smoke Abatement League of Cincinnati, Matthew Nelson, even announced that smoke caused crime. "To say that many of the crimes prevalent in our large cities are directly due to the smoke in the air may seem farfetched, but nothing is more absolutely true."[16] Many reformers in the progressive era clung to nineteenth-century conceptions concerning the strong relationship between health, beauty, and morality. For these reformers, health, cleanliness, and beauty all affected morality. One Milwaukee housewife noted the serious smoke problem in her neighborhood, focusing on the omnipresent dirt. "I sometimes think that it will unchristianize the nation," she said, "and that we will have to recall our missionaries and put them to work at home."[17]

Although smoke severely affected health, aesthetics, and, at least in the minds of progressives, urban morality, the most concrete evidence as to the negative impact of smoke on cities came from economics. In numberless ways smoke taxed urban residents, making their lives not just shorter and duller but more expensive and less efficient. All told, black smoke was an economic disaster.

Filthy soot meant extra expenses in cleaning. Clothes required more frequent washings, and sometimes repeated washings when they became soiled while hanging on the line to dry. The interiors of buildings in smoky cities required extra cleaning, as carpets, furniture, paintings, walls, windows, and everything else with an exposed surface gathered the grimy residue. In libraries, soot damaged books, as what might appear to be years of dust accumulated in weeks. Exterior walls of buildings, as already mentioned, also gathered soot, which if not removed could permanently damage the stone. In Chicago, residents complained about the sticky soot that clung to the white glazed terra cotta exteriors of new skyscrapers. The Chicago *Tribune* claimed that the operators of the loop's giant office buildings spent millions in cleaning costs each year.[18]

On top of cleaning expenses, smoke damage also forced the premature replacement of drapes, carpets, furniture, and the like. Residents of smoky cities often remarked that these replacements often came in dark hues so that they might hide the soot longer than the originals. Pittsburgh became famous for the drab colors of its clothing and furnishings, whose dark hues signaled a surrender to the soot. Buildings required more paint, as well, on exterior and interior walls, and more frequently in smoky cities residents chose dark-

er shades to hide the dirt. Studies also showed that sulfuric acid in soot cor-
roded unprotected iron and steel and ate into building stone, leaving it
porous and weakened.[19]

Smoke also extensively damaged dry goods in stores before they could be
sold. Retailers often complained about smoke-damaged goods, which could
mean annual losses of thousands of dollars in large department stores. One
wholesaler in St. Louis complained that "all classes of merchandise that are
openly exposed for sale deteriorate in appearance and salability from flying
particles of soot and from discoloration and stain incident to smoke." Fine
fabrics, including silk, lace, and ribbon were particularly susceptible to
smoke damage, as were all white goods. To avoid excessive losses due to soot,
retailers spent extra effort and money cleaning their stores. One St. Louis sta-
tionery store claimed to employ three boys whose sole duty was to clean soot
from the stock.[20]

In the smokiest cities, the smoke could blot out the sun, making artificial
light necessary, even at midday. Pittsburgh became infamous for dark smoke
clouds which necessitated the use of street lighting. More commonly, retail
stores, offices, factories, apartments—particularly those near central cities—
burned electric or gas lights all day, adding significantly to the energy costs
of operating businesses in smoky cities.[21]

Smoke also suppressed property values, as clean air became a selling
point in the suburbs and dirty air became a reason to flee neighborhoods near
city centers. Smoke from locomotives greatly depressed the value of property near railroad yards and busy tracks. In 1912, a New York court recognized
the seriousness of the problem when it awarded the United States Leasing
and Holding Company a judgment of $18,000 from the New York Central
Railroad in compensation for the damage the smoke had caused to the leasing price of its property.[22]

While some urbanites feared that efforts to control smoke would prevent
industry from locating in their cities or force existing industry to move to
more hospitable locales, the continuation of smoky conditions most likely
caused greater damage to the economic base of cities than would have abatement's effect on industry. As the Mellon Institute's study of Pittsburgh indicated, smoke had greatly affected the industrial composition of that city. For
the study John O'Connor, a Pittsburgh University economist, identified 264
classifications of industry, of which 245 operated in Pennsylvania in 1909.
Philadelphia had 211 of these industries, while Pittsburgh had only 136.
O'Connor argued that the smoke of the city had retarded the growth of several important industries in Pittsburgh, including several associated with

textiles. The economy of the considerably cleaner Philadelphia was much less adversely affected by smoke.[23]

Obviously, then, businesses experienced significant losses due to coal smoke. They bore the cost of cleaning, painting, artificial lighting, and replacing damaged goods, just as residents did in their homes. But businesses also bore another significant cost due to smoke: the cost of making smoke. As the literature of the day made clear, the creation of smoke wasted coal. The more carbon that went up the stack, the less heat the coal produced. In 1909, Herbert M. Wilson, the chief engineer of the United States Geological Survey and a devoted student of the smoke problem, estimated that smoke represented a loss of 8 percent of the coal burned. Using this estimation, he concluded that the nation wasted 20 million tons of coal each year, at a cost of at least $40 million. As Wilson noted, the heat loss indicated by smoke was not simply equal to the loss of the heat represented in the amount of unburned carbon that left the stack. Smoke indicated an improperly burning fire, either too cool or oxygen-starved. In either case, the presence of smoke meant that the fire could be burning hotter and more efficiently.[24]

Several studies conducted in the early 1900s attempted to put a total price on the smoke damage in American cities. In one of the earliest efforts, the Cleveland Chamber of Commerce conservatively estimated that its city lost $6 million annually to smoke damages. The Chamber of Commerce based its estimate largely on figures supplied by department and dry goods stores, but also considered losses to other businesses, including hotels, hospitals, and banks, as well as to residents directly. Although the committee recognized that smoke also levied a tax on the health of humans, animals, and plants, it did not offer a concrete figure for those costs. Each Cleveland family, the report concluded, lost $44 each year to the tangible effects of the smoke nuisance, the equivalent of four weeks pay for an unskilled worker.[25] In total, the residents of Cleveland paid roughly equal amounts to the "smoke tax" as they did in city taxes. This study became a guide for other estimates. In Chicago, for example, the chief smoke inspector, Paul Bird, assumed that his city was one-third less smoky than Cleveland and set the loss at $8 per capita, or a total of $17.6 million annually. In Cincinnati smoke inspector Matthew Nelson estimated an annual loss of $8 million in his city. In 1909, Herbert Wilson estimated that the United States lost $500 million in total due to smoke damages.[26]

Although the figures Wilson, Nelson, and Bird publicized were little more than guesses, they constituted conservative estimates of an obviously costly problem.[27] In the end, however, no one could propose an economic figure for the true cost of smoke in American cities. No one offered figures for the cost

of shortened lives, chronic unhealthfulness, the drudgery of constant cleaning, or the dullness of dreary days. Nor could municipal leaders offer an estimate for the value of their cities' reputations. Many urbanites feared that
cities of soiled repute would suffer economically, and even in prosperous
cities like Chicago and Pittsburgh, residents suspected that their filthy reputation adversely impacted the local economy. In 1891, a former chief inspector of the Chicago Health Department, Andrew Young, declared that "[a] bad
reputation for polluted water supplies, dirty streets, or impure air is of a direct and continual menace to the city's prosperity and progress."[28]

At the same time, residents of cities with cleaner air, such as New York,
Boston, and Philadelphia, credited their pure atmospheres with aiding economic growth. As Andrew Carnegie noted in 1898, New York's reputation as
a fine residential city continued to attract the nation's wealthy and thereby
the nation's wealth. Carnegie himself chose to live far from his smoky Pittsburgh firm in a comfortable and clean Fifth Avenue mansion. "I have just
spent ten days in Pittsburgh," Carnegie wrote to the New York *Tribune* in
1898, "and smoke is the only impediment to the future of that city." As for his
adopted home he claimed, "There is no question so vital to the interests of
New-York as a residential city, attracting prominent people of other States to
it, as the question of smoke." The *Tribune* agreed, arguing that due to increasing smokiness, "We are on the verge of losing our most valuable pos-
session," and that, "Reputation, honor, comfort, health, prosperity are all at
stake." Meanwhile, even as Pittsburgh's economy continued to boom, its
reputation continued to suffer. While Carnegie could flee the smoke clouds,
the creation of which he played no small role in, others could not. One wonders what the residents of Chicago thought about the Appalachian smoky
city when they read in their *Record-Herald*, "City Grimy, Bride Ends Life; Girl-
Wife Unable to Stand Smoky Pittsburgh, Wanted Cleveland." Surely that
city's reputation could get no lower.[29]

Although this bride's reaction to Pittsburgh's smoke was unusual, inas-
much as studies showed the nation's smokiest cities had no higher rate of suicide than did its cleanest, the irrationality of her reaction to the smoke clouds
was not atypical. Average residents' reactions to foul air tended to be vis-cer-
al, and tended not to center on figures concerning this or that cost of cleaning soot. The dirt and darkness elicited emotion as much as they did thought,
and many residents of smoky cities expressed concern over the psychological effects of the dark clouds. J. E. Wallace Wallin, the director of the Psychological Clinic at the University of Pittsburgh, authored a bulletin on the subject for the Mellon Institute's investigation. While Wallin described the

"chronic ennui" and depression he found in Pittsburgh's residents, he also described the effect of smoke on his own work. Having been in Pittsburgh for less than two years, Wallin had not yet grown accustomed to the dirty atmosphere. "Clear, trenchant, reflective thinking seems to have been more difficult," he wrote, "and the attempt to write concisely, incisively and perspicuously has seemed more labored." Smoke hung heavy in the consciousness of Wallin, as it did in other city residents.[30]

Fiction writers knew smoke carried significant meaning for urban readers, and they used descriptions of smoke-darkened atmospheres with great effect. Upton Sinclair made clear the emotional power of smoke in his celebrated 1906 work *The Jungle*. As the novel's working-class protagonist, Jurgis Rudkus, looked upon Chicago's packingtown for the first time, the smoke caught his eyes as much as the stench filled his nostrils. Jurgis looked upon "half a dozen chimneys, tall as the tallest of buildings, touching the very sky—and leaping from them half a dozen columns of smoke, thick, oily, and black as night." The smoke united in the sky to form "one giant river." This was conspiratorial smoke, gathering in an oppressive show of strength. This was smoke that "might have come from the center of the world," from hell itself. It did more than darken the sky; it threatened all who observed its "writhing." Sinclair's smoke was foreboding at best, evil at worst. If the smoke represented jobs for immigrants, Jurgis among them, it also represented the horrifying conditions under which they labored. If Jurgis could not suspect the meaning of the writhing smoke clouds, Sinclair knew his middle-class readers would. Here in the clouds hung fear and uncertainty. Here in the clouds were the contradictions of industrial civilization.[31]

Smoke so thoroughly vitiated American industrial cities that few commentators offered a complete representation of its total effect. Physicians commented on the health aspect. Engineers commented on the waste of coal represented in smoke. Women tended to focus on the aesthetic aspect, and on the extra cleaning the soot required. At certain moments, however, a more complete picture of the gravity of the smoke problem became clearer. One such moment occurred in Chicago in 1907 as the City Club's smoke committee conducted a study of the "smoke evil." In one week, city planner Daniel Burnham, famed for his work on the White City of the 1893 World's Fair and for his co-authorship of the 1906 *Plan of Chicago*, made clear the importance of smoke abatement to beautifying the city; the nation's premier retailer, Marshall Field, noted that the soot tax on his business, the famed Chicago department store that bore his name, was greater than the real estate tax on his property;[32] a physician from the Chicago Tuberculosis Institute emphasized the "ravages of smoke on the vital organs," and declared that smoke aided in the spread of consumption; and Lester Breckenridge, a University of Illinois mechanical engineer then engaged in a major study of Illinois coal, declared positively that smoke was unnecessary. The Chicago *Record-Herald*, long active in the smoke abatement campaign, made certain to include all these voices in the daily paper.[33]

Despite the seriousness of smoke's myriad effects on cities, urban residents who worked for decades to find solutions to the growing air pollution crisis never fully succeeded. In fact, of the major pollution issues facing rapidly industrializing cities, smoke received the least effective attention. Most municipalities found adequate, though temporary, solutions to their garbage, sewage, and water supply problems, even as smoke continued to envelop their cities. More than one commentator noticed the inconsistency with which cities worked to solve their environmental problems. "With the development of urban systems," one author wrote in 1907, "the providing of adequate water supplies, the improvement of transportation facilities, the humanization of urban conditions, the atmosphere of the great city has been converted into a deleterious vapor." Did urban residents place greater value on pure water, sewage removal, and garbage collection than on pure air, or did the smoke problem just prove more difficult to solve?[34]

The lamentations of thousands of city residents concerning smoke suggest no lack of appreciation for pure air, clear skies, and sootless surroundings, and, on the surface at least, the proposition that cities simply found the smoke problem more difficult to solve seems apparent. Obviously something about the smoke problem prevented an easy solution. But, when we consider the expense and effort some cities endured to correct other environmental problems, this conclusion becomes less tenable. In the mid-1800s New York City, for example, committed decades of work and millions of dollars to the creation of the Croton Reservoir and its aqueduct system to ensure that ample potable water would reach Manhattan. At the same time, Chicago elevated all of its streets and constructed a drainage canal to keep sewage from stagnating in city streets, Lake Michigan, or the Chicago River. Thus other environmental problems impelled expensive and technically difficult solutions, suggesting that cities did have the patience, desire, and financial wherewithal to solve complex environmental dilemmas.[35]

Two other factors were more important in differentiating the smoke problem from other environmental crises of the day. First, unlike the solutions concerning sewage, garbage, and pure water, those offered for the smoke problem all suggested some sacrifice from industry. While sewage, water, and garbage were urban problems, city residents defined smoke largely as an industrial problem.[36] Few urbanites, even those most active in the anti-smoke campaign, proved willing to require the large sacrifices from business which seemed necessary for a real cessation of smoke. Thus, while solutions to garbage, sewage, and water problems aided businesses just as they did city residents, solutions to the smoke problem were potentially threatening to the industrial order, or at least so they seemed. Although creating smoke carried its own costs for polluters, in loss of fuel, some businesses feared that the costs of control would be greater, particularly if government mandated immediate abatement or specific control devices. Many businessmen expressed concern about the expense of new equipment and the very real possibility that the new installations would not effectively control smoke.

Second, while urban residents could express equal appreciation for pure air, pure water, and clean streets, they did not place all environmental problems in the same category. In the late 1800s the dominant miasma theory of disease held that foul odors, emanating from sewage or some rotting organic matter, and gases discharged from stagnant waters, caused specific diseases and general unhealthfulness. As the century progressed, more people came to accept the germ theory of disease, which asserted that microbes, bacteria, caused diseases, with or without the presence of foul water and odors. Both of these theories, and the odd combinations of the two which city residents held during the long transition, suggested the extreme importance to health of adequate pure water supplies, efficient and timely removal of garbage and offal, and a properly draining sewage system. Perhaps more important to municipal action on these issues than their relationships to diseases were the types of diseases they seemed to cause and how those diseases struck the city. By the 1850s, city residents associated frightful epidemics of cholera and typhoid fever, both waterborne diseases, with poor sewerage, dirty streets, and impure water sources. The crises wrought by epidemics, which could kill thousands of city residents in a matter of weeks, impelled action.[37]

Unlike the consequences of impure water and inadequate sewerage and garbage removal, the problems associated with smoke were not acute. Smoke caused no epidemics, no brief, intense crises.[38] The problems associated with smoke were endemic, not epidemic, and therefore controlling smoke was less likely to receive the political energy and the public willingness to sacrifice which were required for its success. As the editor of *American Medicine* summarized in 1902, "Probably one of the chief reasons for the indifference to the reform of the smoke nuisance is that most people think of the matter not as one of health but of esthetics." As long as residents thought of smoke as only an aesthetic problem, palls would persist, since most urbanites could endure ugliness in the pursuit of economic progress.[39]

To impel action on the smoke issue, antismoke activists would have to convince fellow residents that the endemic problems caused by smoke required immediate action, that smoke had just as serious implications for health and morality as did foul water and rotting garbage. Antismoke activists—women, physicians, municipal and mechanical engineers, businessmen, artists, and academics among them—would have to convince fellow residents that smoke was an environmental problem on the scale of inadequate or fouled water supplies, stagnating sewage, or rotting, accumulating garbage. Reformers would have to convince their fellow urbanites that smoke posed a real threat to their society, that health, beauty, and morality were as important as wealth and growth, and that a civilized nation required civilized air.

Trouble in the Air
The Movement Begins

> Our once pellucid and crystalline atmosphere is obscured;
> buildings are befouled within and disfigured without; the beau-
> ty of the city and of its prospects is marred; the comfort of the
> people is impaired, and the health of the people is menaced.
> Those are the troubles that are in the air of New-York, and that
> are transforming the air from a pure delight into an incubus.
> —New York *Tribune*, 16 September 1899

As Americans turned to coal to fuel their industry, heat their homes, and drive their transportation, smoke thickened over the nation's cities. Through most of the nineteenth century, urbanites tended to accept the smoke, only occasionally expressing discomfort and discontent, and rarely demanding relief. Many urban residents even articulated an appreciation for the smoke or at least acceptance of the pollution as a necessary part of industrial progress. But in the last decade of the 1800s, perceptions of smoke began to change and formerly mild protestations turned anxious and intense. According to many middle-class reformers, the smoke clouds that had slowly enveloped America's industrial cities suddenly posed a threat to their existence. The organized effort to abate smoke began in the 1890s, as middle-class urbanites reappraised their industrial civilization and offered a new vision for the urban environment, including its fouled air.

Of course the reformers of the 1890s were not the first to notice the smoke or issue complaints. After decades of worsening air quality many urbanites had identified a multitude of smoke-related problems. In Pittsburgh, the American city where soft coal first played an important economic role, objections rose with the smoke. One Pittsburgher expressed his dismay at the smoke pall in a letter to the Pittsburgh *Gazette* as early as 1823. While appreciating the prosperity that coal and industry had brought to his city, the writer noted that "at the same time the increased number of chimneys, pouring forth dark and massive columns of smoke, begins to be felt as an almost intolerable nuisance." Decades later Pittsburgh's dirty reputation had spread, aided by reports from visitors. One such visitor, James Parton, a contributor to the *Atlantic Monthly*, described Pittsburgh to the nation in 1868: "The town lies low, as at the bottom of an excavation, just visible through the mingled smoke and mist, and every object in it is black. Smoke, smoke, smoke,—everywhere smoke!" Sixteen years later, when Willard Glazier portrayed Pittsburgh in his travel book *Peculiarities of American Cities*, he also emphasized the serious aesthetic and psychological implications of the dense smoke. "In truth," Glazier wrote, "Pittsburgh is a smoky, dismal city, at her best. At her worst, nothing darker, dingier or more dispiriting can be imagined."[1]

Although Pittsburgh became infamous for its smoke pall, it was not the nation's only smoky city. In Cleveland, where coal use grew rapidly after the Civil War, an editorial in the *Daily Leader* sounded a warning in 1869: "Cleveland is in danger of losing its reputation as a beautiful city, for the same cause which has made Pittsburgh proverbial as the dirtiest of western cities—the growth of its manufactures, and the use of bituminous coal therein, clouding the city with smoke." While the *Daily Leader* praised the economic growth that fouled the air, it listed several serious threats posed by the smoke, including "constant soiling of clothing," "unhealthiness both to animal and vegetable creation," and a total economic damage of up to a million dollars annually.[2]

Several of the nation's young industrial cities, including Pittsburgh, Chicago, Cincinnati, and Milwaukee, witnessed attempts to abate smoke emissions well before the 1890s. These early encounters produced few tangible results, however. Many early antismoke efforts concentrated on individual offensive smokestacks, rather than smoke generally, and the struggles tended to be short-lived. Even when they culminated in the passage of broad antismoke ordinances, these efforts did not signal the coming of cleaner air. Although urban residents did force the improvement of some smoking stacks through narrowly focused campaigns, the booming industrial economy brought several new, dirty chimneys to replace each reformed stack.[3]

Some cities did attempt to legislate against smoke emissions before the 1890s. In 1869, Pittsburgh passed an ordinance forbidding the use of soft coal in locomotives within the city. Although the city council mustered the support to pass the legislation, limited public backing and an absolute lack of enforcement made the ordinance meaningless. Few Pittsburghers could envision their city without soft coal, and restricting its use did not prove an effective means of controlling smoke.[4] In 1881, Dr. Julia Carpenter, a prominent Cincinnati physician, gathered support for a new ordinance, to replace an ineffective 1871 law. While the new legislation provided for the appointment of a smoke inspector, neither the 1881 ordinance nor its 1883 replacement proved of much value. Reformers failed to form an organization to exert pressure on the city, and the ordinance remained largely unenforced. Chicago, too, passed antismoke legislation in 1881, but even with judicial support the ordinance could not prevent a continuing decline in air quality.[5]

In most cities efforts to control smoke did not begin with municipal legislation but with narrow public attacks on individual polluters. Although concentrated activism by citizens could force smoky businesses to attempt abatement, given the limitations of available technology, satisfactory cessation of smoke nuisances often required offending businesses to suspend operations or relocate. In Milwaukee, for example, residents of the seventh ward took their complaints directly to the Steam Supply Company when that firm's smoke emissions became offensive in the fall of 1879. Aided by the support of the city health department and the *Daily Sentinel,* the residents' activism compelled the company to install three new furnaces and a "smoke consuming apparatus" to alleviate the problem. The smoke consumer proved ineffective, however, and the company removed it, forcing Health Commissioner Orlando Wight to issue an order for the "immediate abatement of the nuisance." But despite the threat of municipal interference, the Steam Supply Company could find no easy solution to its smoke problem, and two years later the stacks continued to issue smoke. The neighborhood activism had drawn attention to the smoke problem, but it had not impelled a solution.[6]

In the many instances when public pressure failed to bring relief from smoke nuisances, residents near large polluters could seek justice in the courts. Before the 1890s, however, judges tended to show little sympathy for smoke complainants. In 1871 the Pennsylvania Supreme Court, found against a plaintiff who had hoped the justices would prevent a brick factory from issuing the smoke that damaged his vineyards and orchards. In the opinion, Justice Agnew argued that brick making was a necessary employment near a large city, in this case Pittsburgh, and that the court could not interfere in such a beneficial enterprise. "The people who live in such a city or within its sphere of influence do so of choice," Agnew wrote, "and they voluntarily subject themselves to its peculiarities and its discomforts, for the greater benefit they think they derive from their residence or their business there."[7]

Agnew's reasoning found broad support in Gilded Age jurisprudence. In 1876, a New York judge asserted, "If one lives in the city he must expect to suffer the dirt, smoke, noisome odors, noise and confusion incident to city life." In 1880, a Kentucky judge echoed this argument while finding in favor of a coffin factory in Louisville which produced heavy smoke. Chief Justice Pryor argued that manufacturing interests were "necessary and indispensable to the growth and prosperity of every city," and even though they marred the cleanliness and beauty of surrounding neighborhoods, the "individual comfort must yield to the public good." Essentially these Gilded Age courts ruled that a citizen's right to clean air varied by the type of community in which the person lived. City dwellers could not enjoy the same right to a clean and healthful environment that rural residents expected. In 1868, the Pennsylvania Supreme Court made clear the distinction in rights, noting that every citizen had a right to "pure and wholesome air, at least as pure as it may be, consistent with the compact nature of the community in which he lives."[8]

In the two decades after the Civil War, then, courts offered little protection from urban air pollution, as judges endeavored to protect economic growth. There were important exceptions, however. Some businesses did produce such obvious nuisances in urban settings that courts often found in favor of suffering plaintiffs. Smelting firms, for example, produced air pollution recognized to be qualitatively different from most industrial smoke. The lead and arsenic vapors issued from the Pennsylvania Lead Company's smelter outside Pittsburgh convinced the state supreme court to rule in favor of a small land-owning plaintiff and against the producer of one-fifth of the nation's entire lead supply. In the opinion, Justice Stowe wrote, "To undertake the business of lead smelting in the midst of a rich suburban valley, occupied by farms and country residences, was, to say the least of it, not very prudent." Ten years later a Michigan court found similarly in a case against the Detroit White Lead Works, which produced offensive lead-laden smoke and foul odors. Neighbors complained of nausea, headaches, and vomiting, convincing the court to find against the defendant. Thus, when plaintiffs successfully argued that offending air pollutants caused significant health problems, they could expect real relief via court injunction. The negative health effects of chemicals containing lead and arsenic made smelters obvious targets for such action.[9]

Before the turn of the century, the link between coal smoke and health problems remained tenuous, and unlike producers of poisonous lead smoke, businesses yielding largely carbonous emissions often earned the protection of courts. Judges regularly ruled that mean inconveniences, such as dirt and discomfort, did not warrant court interference in legitimate economic activity. Again, however, there was one major exception: the courts often found in favor of plaintiffs hoping to expel polluting industries from residential neighborhoods, even when the complainants made no special plea concerning the health implications of the pollution. In this pre-zoning era, middle-class residents used the courts to protect the integrity of their communities. Some urbanites even took businesses to court before they began operations in their neighborhoods, thereby complaining of a nuisance that did not yet exist. In many cases, courts found in favor of complainants who could argue that the polluting industry was out of place. In 1895, for example, nineteen residents of a Detroit neighborhood argued successfully against a small forge that had begun operations near their homes. The judge considered the character of the neighborhood, and no doubt of the middle-class complainants as well, in finding the smoke and noise from the forge a nuisance and demanding its removal.[10]

In the decades before the 1890s, then, efforts to control emissions into the urban atmosphere yielded some tangible results, particularly in banishing many toxic polluters from densely populated areas and in protecting some residential neighborhoods from encroaching industry. But these piecemeal efforts could not produce citywide improvements, and the air quality of the nation's industrial cities continued to deteriorate.

The form of antismoke activism began to change, however, in the 1890s. Private organizations began studying the smoke nuisance and lobbying for legislation focusing specifically on the emission of black smoke. Several types of interest groups took up the fight against smoke, including health protective associations, women's social clubs, businessmen's clubs, and even organizations formed specifically to promote the abatement of smoke. Pittsburgh, St. Louis, and Cleveland all witnessed the initiation of organized antismoke activism in 1892. Reformers in these cities moved beyond attacks on individual polluting stacks and created standing bodies to investigate smoke as a citywide problem that demanded a citywide solution.[11]

In the 1880s Pittsburgh experienced a respite from its infamous smoke, when a reliable supply of natural gas greatly reduced the amount of coal consumed in the city, perhaps by as much as two-thirds. But by 1890 coal consumption again soared, as the supply of the clean-burning natural gas began to falter. "We are going back to the smoke," lamented one Pittsburgher in 1892. But the experience with relatively clean air had convinced many city residents not to accept a return to a smoky atmosphere. In 1891, the Women's Health

Protective Association of Allegheny County, an organization of middle-class women, many of them wives of influential men in Pittsburgh, took up the fight against the returning smoke. In that year the association's secretary, Mrs. Imogene Oakley, the wife of a prominent broker in the city, researched the engineering aspect of the smoke problem in an effort to discover the latest and most effective "smoke consuming" devices. She turned to Chicago's Citizen's Association, which had listed such devices in an 1889 report. Oakley and other members of the Health Protective Association, armed with this new information, began working the following year with the Engineers' Society of Western Pennsylvania, which had created a committee on smoke prevention in reply to the women's pressure. The women also lobbied the city council for the creation of a new, effective ordinance.[12]

Although these women understood the need for a technical solution to the smoke problem, they defined the problem itself in health and aesthetic terms. The women turned to physicians and each other as experts on the smoke problem, for they knew no other groups in society who so well understood the effects of smoke on households and health. The title of their organization, the Women's Health Protective Association, underscores the members' belief that women had a special responsibility to protect the health of their families, and by extension, the health of all their city's residents. Middle-class, Victorian society held women, working in the private sphere of home and family, as the traditional protectors of health, cleanliness, and morality. When middle-class women began to organize, particularly into social clubs and narrowly focused interest groups such as the Health Protective Association, they addressed many environmental issues that directly affected health, cleanliness, and aesthetics. These women reformers also assumed that ugly, dirty, and unhealthful environments affected the morality of city residents. In addition to smoke, many women's organizations worked on other sanitation issues, including garbage collection and street cleaning, two problems that the Pittsburgh women had tackled since the founding of their league in 1889. Through these organizations active women extended their spheres from the home to the public and acted as "municipal housekeepers." In the process, they ensured that the public discourse of environmental problems would center upon health and aesthetics, and that the dialogue would contain a sense of moral imperative.[13]

Those who feared the fledgling smoke abatement movement in Pittsburgh, suspecting that it would have a detrimental impact on the city's industry, directly confronted the women's assertions concerning the health effects of smoke. In a presentation concerning smoke to the Engineers' Society of Western Pennsylvania, long-time member William Metcalf responded directly to the charges of the women who attended his speech. "I assert that there is nothing particularly unhealthy about smoke; on the contrary, it may mitigate other and worse evils." During the discussion of Metcalf's presentation, a physician in the crowd, in attendance at the behest of the women, used an article from the *British Medical Journal* and his own logic to conclude that smoke was indeed harmful to health. "Take an atmosphere . . . in which trees will not grow," said Dr. Sutton, "and it is not a good atmosphere for man, and such is the atmosphere of the city of Pittsburgh." In response, William P. Rend, a visiting Chicago coal dealer who had been fighting against the nascent antismoke movement in his own city, defended smoke. "Now, I am not a doctor, but if I was, I probably would differ with that gentleman," began Rend. "I believe that smoke is healthy. I challenge the doctor to prove that it is unhealthy."[14]

Despite the opposition expressed during these discussions, the Engineers' Society formed a committee to study the technical aspects of smoke and expressed official support for the Health Protective Association and for the passage of a new ordinance. The city responded quickly to the pressure of the women and the engineers and in early 1892 passed a law prohibiting any chimney or smokestack associated with a stationary boiler to emit bituminous coal smoke. This ordinance had severe limitations, however. Aside from not addressing locomotive smoke at all, it applied to only stationary boilers within a certain district, and the boundaries for the district neatly skirted Pittsburgh's industrial centers. The city council drew irregular boundaries specifically to exempt iron and steel works and other heavy industry. The city had once again created an ordinance that could not abate the smoke pall.[15]

In the same year that the Women's Health Protective Association initiated Pittsburgh's antismoke crusade, in St. Louis a social club of prominent women, the Wednesday Club, helped create and support that city's new antismoke organization, the Citizens' Smoke Abatement Association. The Wednesday Club, through a committee established for the cause, joined the Citizens' Association and the Engineers' Club of St. Louis in lobbying for a new antismoke ordinance. The women passed a resolution announcing their support for smoke abatement and the reasoning behind their effort: "We feel that the present condition of our city, enveloped in a continual cloud of smoke, endangers the health of our families, especially those of weak lungs and delicate throats, impairs the eyesight of our school children, and adds infinitely to our labors and our expenses as housekeepers, and is a nuisance no longer to be borne with submission." After listing the serious effects of smoke on city residents, the women declared the need to unite in "denouncing this condition of our otherwise beautiful city, and in protesting against its continuance." The women resolved to support the Citizens' Smoke Abatement Association and to aid in the passage of a new smoke ordinance. In 1893, they succeeded in this latter goal, as the city passed legislation which declared "the emission into the open air of dense black or thick gray smoke" a nuisance and created a three-person commission to enforce the new law. Unlike Pittsburgh's ordinance, the St. Louis law had a significant effect on the air quality of that city over the next three years, as the smoke commissioners educated boiler room operators in proper firing and convinced proprietors of smoke prevention's economic advantages. In 1896, however, a successful legal challenge removed the law from the books, as the state supreme court determined the city had no authority to declare smoke a nuisance.[16]

Just as in Pittsburgh and St. Louis, Cleveland's organized antismoke movement also began in earnest in 1892. Although Cleveland had passed a smoke abatement ordinance in 1882, the city accomplished little in the way of reducing smoke until Charles F. Olney led the creation of the Society for the Promotion of Atmospheric Purity ten years later. Olney, an art teacher and owner of an art gallery, served as the Society's president and directed the fight for clean air largely on aesthetic grounds. "Let our avenues, streets and parks but shine with beauty," he said before a gathering of the Fine Arts Club; "let the breezes of Heaven but reach their destination uncontaminated by man." Olney's Society, which claimed four hundred of the city's prominent citizens as members, aided the city in the enforcement of its new ordinance, which had empowered the health department to abate smoke nuisances. The law prohibited the emission of dense smoke "from anywhere in the city, or from the smokestack of any boat, locomotive or stationary engine or boiler within the limits of the city of Cleveland." As in St. Louis, the Cleveland law was much more inclusive than Pittsburgh's legislation, and the new ordinance proved to be too sweeping in its regulation and too vague in its wording. In 1896 an appeals court declared the ordinance void while overturning the first conviction of a smoke offender under the law.[17]

In all three cities, even though different types of organizations led the crusade against smoke, the most active reformers defined the problem in distinctly feminine terms.[18] This feminine definition involved four interrelated facets: health, aesthetics, cleanliness, and morality. Early antismoke activists expressed deep concern for the beauty of their cities, often commenting on smoke's obstruction of scenic views. But generally when reformers spoke out against the smoke, they connected smoke's obvious aesthetic problems with more tangible negative effects, including extreme filth from constant soot fall. That coal smoke was a cleanliness problem could not have been more plain. In effect, smoke and soot violated the sanctity of the home, a transgression that remained central to middle-class women's interest in the issue. As one Milwaukee housewife remarked in 1903, "The smoke blackens the entire house and makes it look dingy, inside and out." Indeed, the dark and filthy quality of soot lent itself to the lexicon of antismoke reformers. Smoke was "impure," it "fouled" and "defiled."[19]

But reformers could not rely solely on cleanliness or aesthetic issues to force municipal action. Instead, the initiation of effective government regulation of smoke would require more serious implications than soiled furniture and spoiled views. Thus, employing arguments which had impelled municipal action toward improvements in sewerage and water supplies, smoke abatement reformers related air pollution to public health issues. Activists eagerly connected smoke with a wide variety of health problems, arguing that it had ramifications as various as acne and lung disease. Indeed, reformers could even attribute conflicting symptoms to smoke poisoning, such as diarrhea and constipation. Lung ailments, however, received most attention from antismoke reformers, who regularly attributed high tuberculosis death rates to filthy urban air, and connected whooping cough, bronchitis, and pneumonia to smoky atmospheres. Reformers also attributed psychological ailments to smoke, including a vaguely described malaise caused by the sapping of energy, and a depression caused by decreased sunlight, which some commentators even suggested led to increased suicide rates in smoky cities. These psychological implications could also conflict, however, as smoke received blame for inciting both lethargy and criminal activity.[20]

Since progressive-era reformers tended to connect beauty, cleanliness, and health to moral development, they also argued that smoke adversely affected the morality of urban citizens. Excessive dirt, the popular argument ran, created a citizenry blind to the virtues of cleanliness and incapable of maintaining healthful homes. As Mrs. John B. Sherwood, president of Chicago's Women's Club, concluded, "Chicago's black pall of smoke, which obscures the sun and makes the city dark and cheerless, is responsible for most of the low, sordid murders and other crimes within its limits. A dirty city is an immoral city, because dirt breeds immorality. Smoke and soot are therefore immoral."[21] Like other middle-class environmental and social reform efforts, antismoke activism could smack of paternalism, in that reformers thought their efforts to clean working-class neighborhoods and tenement slums would improve the morality of the lower classes. Still, most antismoke activists were much more concerned with the air pollution's effect on their own homes and neighborhoods, and moral arguments received much less attention than those concerning cleanliness and health.[22]

The definition of smoke articulated by middle-class women found wide acceptance outside the lay organizations dedicated to smoke abatement. Even engineers interested in controlling smoke tended to define the problem in the terms stressed by lay reformers, at least in the early years of the movement. The St. Louis Engineers' Club, for example, reported in 1892 that smoke was partly an aesthetic problem, since smoke reduced the impulse to ornament the city, and partly a health problem, as witnessed by the prevalence of lung and throat troubles. These engineers also noted a moral loss suffered by "those who per force endure dirt and unwholesomeness." Certainly all engineers recognized that smoke control would require some scientific solution. And although they discussed potential solutions in mechanical, scientific terms, many engineers described the problem itself in rather unscientific ways. For some engineers the problem remained intangible and unquantifiable, but still very real and in need of urgent attention. By the turn of the century, then, many engineers had accepted the dominant definition of the smoke problem established by lay reformers.[23]

Of the four facets of this feminine definition, the one relating smoke to public health was far and away the most important. By attributing a myriad of negative health effects to smoke, reformers convinced city officials that air pollution, like the issues of pure water, sewage, and garbage disposal, required immediate municipal action. In the late 1890s, editorials that appeared in the Christian reform magazine the *Outlook* made clear the importance of relating smoke to health. As soft coal continued to make inroads in eastern industrial markets, the *Outlook* warned of a gathering smoke cloud in New York City, emphasizing the importance of aesthetics and admonishing that "God did not make the world beautiful as a matter of personal caprice." Rather, as the editors concluded elsewhere, "The sky is a gift of God which it is a profanation to defile." Clearly the *Outlook*'s primary concern for smoke lay in its aesthetic effects. The magazine also frequently mentioned the health effects of impure air, however, without ever offering specifics. Showing an obvious understanding of the situation, the magazine argued, "The soft-coal industry is one of immense magnitude, and could be interfered with only on the ground of public health." In other words, those acting to protect the purity of their atmosphere were more likely to succeed if they argued in terms of health rather than in terms of aesthetics, even if they were more certain of the effects on beauty.[24]

Smoke's implication in matters of health helped create the imperative for action that sparked local antismoke movements throughout industrial America, even before research proved that the need for relief was real. In asserting that smoke caused serious health problems, early reformers were not simply publicizing accepted knowledge. In fact, by connecting smoke to unhealthfulness in unprecedented ways reformers hoped to change public perceptions of smoke. For centuries physicians and laymen had attributed antiseptic qualities to smoke, and healers had long used smoky coal to "purify" disease-ridden air. Even into the late 1800s and early 1900s, some Americans expressed doubts about the unhealthfulness of smoke. In 1897, during the Franklin Institute's extensive review of the smoke problem, a Cornell University engineer commented that smoke "is rather healthful than otherwise." Another defender of smoke in Philadelphia claimed that the purifying, disinfecting qualities of smoke had reduced the incidence of malaria in the Juniata Valley, where passing Pennsylvania Railroad locomotives issued regular salubrious emanations from their stacks.[25]

Although some still held to this old conception of smoke as disinfectant, by the turn of the century most students of air pollution understood that it was unhealthful.[26] Even in 1897, at the Franklin Institute's discussion of smoke, mention of the healthful effects of smoke met with little support and caused considerable amusement. By the 1890s, many physicians had come to believe that smoke had a deleterious effect on health, even though they had very little scientific data to confirm their suspicions. These physicians expressed certainty as to the negative effect of smoke on health, but uncertainty as to exactly how smoke affected humans.[27]

The unhealthfulness of foul air was certainly not a new issue, but whereas urban air had been a health concern for decades, coal smoke had little to do with early fears, leaving antismoke reformers in the 1890s with little medical foundation from which to build. Early concerns for urban air quality centered upon "vitiated air," which sanitarian John Griscom discussed in detail in *The Uses and Abuses of Air*, published in 1848. Griscom expressed more concern for stagnant indoor air than he did for outdoor air laden with smoke. Like other sanitarians, Griscom was particularly concerned with school rooms and basement tenement apartments, which he assumed became dangerously low in oxygen due to lack of ventilation. Of course smoke pollution garnered very little concern from anyone before the Civil War, but these old fears of "vitiated air" persisted even after smoke became a much more obvious factor in urban air pollution. Ventilation and oxygen levels remained important issues throughout the 1800s and into the early 1900s, often eclipsing concerns about smoky atmospheres.[28]

For decades the greatest fear surrounding coal fires involved their potential effects on indoor rather than outdoor air. Poorly designed and installed equipment leaked smoke and fumes into tenement apartments, where in cold weather windows and doors remained tightly shut. Both residents and reformers expressed concern over the lack of ventilation in coal-heated rooms, fearing that fumes, even from anthracite coal, could overcome residents. In 1868, a Boston surgeon, George Derby, described the symptoms associated with coal fires, including headaches, nausea, and general languor. Derby concluded that these symptoms resulted from the slow poisoning of residents by "carbonic oxide gas," now simply called carbon monoxide. Derby researched the chemistry of burning anthracite and determined that the solution to the problem was complete combustion, which resulted in the creation of carbon dioxide, a harmless gas, rather than carbon monoxide, which Derby knew to be lethal at certain concentrations. Derby also recommended the installation of airtight stoves and pipes to prevent the inadvertent escape of carbon monoxide gas. Derby expressed no concern for the effect of the thousands of coal fires spewing emissions from airtight pipes into the city. In his eyes, the atmosphere apparently provided an acceptable sink for unwanted indoor pollutants.[29]

By the late 1800s, the fear of "vitiated air" had merged with other concerns about crowded tenement district environments. In the cramped places of the city, in basements, alleys, and narrow streets, oxygen levels dropped, so the theory went, impeding moral and physical development and threatening human health. This concern for air quality reflected a kind of claustrophobia, as a fear of crowded, tight places expressed itself in visions of entire communities gasping for oxygen. Some observers saw the real evil in coal fires as the excessive consumption of oxygen, rather than the production of carbonous emissions. According to this theory, humans competed with fires for the precious resource.

Through the last six decades of the nineteenth century, proponents of urban park development also expressed concern for oxygen levels and the "impurity" of city air. Urban park advocates argued that open spaces provided breathing room, not just for city residents but for the city itself. Supporters often called parks "the lungs of the city," claiming that by promoting the movement of air parks could actually purify a fouled atmosphere, even remove disease-causing miasmas. Under the theory of the day, sanitarians identified motion as "a natural method for the purification of the atmosphere." As one physician noted in 1896, "When not in motion, air stagnates as water does and becomes offensive and bad, because it is easily impregnated with fine animal and vegetable dust as well as noxious gases."[30]

In the 1890s, an Englishman, Robert Barr, took the fear of stagnating, oxygen-deprived air to its extreme in a short story, "The Doom of London," in which a man, fortunate to have a machine that produces oxygen, walks dazed through London during a heavy smoke fog. All of the city's residents lie collapsed in the streets, dead from suffocation. Having burned all the life-giving oxygen, the fires of London went out, including those of locomotives at the train station, where thousands had crammed into rail cars hoping to escape the suffocating fog, only to meet their fate as the locomotive fires consumed the city's last molecules of oxygen and then went out.[31]

The fear of oxygen deprivation may have distracted the energies of well-meaning reformers from the more serious air pollution problem, coal smoke. Still, early activists proved just as willing to assume the unhealthfulness of smoke as others were to assume the reality of dangerous oxygen depletion, which actually was far from real. Indeed, early antismoke rhetoric attributed a wide variety of health problems to coal smoke well before scientific evidence could support, or in many cases disprove, such claims. Women who drew general conclusions about the close relationship between cleanliness and health needed no scientific data to support the idea that filthy smoke and soot had deleterious effects on health. And given women's roles as keepers of the house and protectors of morality, female voices decrying the ravages of impure air echoed deep into society.

Around the turn of the century, however, women increasingly lost their public authority in matters of health to professional medical doctors. Although middle-class women often allied themselves with physicians in efforts to improve urban health and sanitation, these two groups also struggled against each other for authority over health issues in the public sphere. As physicians began to engage the smoke problem as a profession, through research and dissemination of information, women lost much of their influence over the issue. In a society increasingly enamored with science, lay opinions on issues of health, including those of middle-class women, gradually lost much of their relevance. Women could identify smoke as filthy and impure, but shortly after the turn of the century only physicians possessed the expertise required to determine its actual effects on health.[32]

American physicians added little to the discussion of smoke and health before the turn of the century, perhaps because they were unwilling to abandon old theories concerning the positive effects of smoke or to draw new conclusions without new evidence. When physicians did enter the early anti-smoke dialogue, they often simply repeated the general arguments against smoke made in other, nonprofessional quarters. Few American physicians conducted research to determine the exact effects of smoke on health, and most studies that did appear were loose epidemiological studies with little scientific value. Often, health officials simply compared the death rates of various cities and estimated levels of smokiness. Using these types of studies, some physicians determined that smoke did not greatly influence mortality, since Pittsburgh and London, both undeniably smoky cities, did not have unusually high death rates. Other studies used more specific data—death rates from tuberculosis, for example—and compared ward-level statistics rather than citywide data. Nevertheless, these studies, too, were so broad as to only suggest that many factors in addition to smoke affected death rates from disease.[33]

In the early 1900s, physicians became much more vocal in attributing health problems to smoke, even before new research provided conclusive evidence. In the spring of 1905, for example, the *Journal of American Medical Association* announced, "It is certain . . . that smoke and other products of soft coal combustion vitiate the atmosphere and constitute an influence genuinely depressing to the organism." Continuing with vague assertions, the article claimed that children who could not escape the smoky atmosphere of city neighborhoods tended "to be pale and flabby." Even while the *Journal* listed these general health effects of smoke, the organization nevertheless remained quite removed from the smoke issue. In fact, in the same article the *Journal* claimed that "the crusade against unnecessary city smoke is more social than medical."[34]

In 1906, however, the position of the American Medical Association began to change. In July of that year its *Journal* reviewed the findings of a German physician, Louis Ascher, who had published his results in Stuttgart the previous year. Ascher had compiled mortality statistics which revealed an increase in acute pulmonary diseases in manufacturing districts, where smoke was particularly thick. He also conducted laboratory studies in which he exposed animals infected with tuberculosis to varying levels of smoke. Ascher found that small mammals died faster from tuberculosis when exposed to smoky atmospheres. In the light of Ascher's findings, the editors of the *Journal* recommended that smoke control "should be made part of the general plan of the campaign now in progress against tuberculosis and pneumonia."[35]

Just a month later, the *Journal* reported the findings of five participants in a discussion of smoke and health sponsored by the Philadelphia County Medical Society. Those physicians emphasized the influence of smoke on nose, throat, and eye irritation and the increased potential for infection of irritated organs. In 1907, two American physicians published influential articles on the certainty of health effects of smoke. Dr. Theodore Schaefer of Kansas City reported the results of his research concerning the health effects of sulphur dioxide, and Dr. Abraham Jacobi of New York offered a more general summary of smoke's relationship to health. Both men relied heavily on European research conducted over the previous decade, particularly in Germany and England, and both men concluded that smoke posed a serious health threat to urban residents and recommended government regulation of emissions.[36]

Physicians also began to take more seriously the indirect effects of smoke, particularly on ventilation and sunlight. Some physicians expressed a fear that tenement residents kept windows closed during particularly heavy weather, with the intention of preventing soot from gathering in their apartments, but with the inadvertent effect of keeping "vitiated" air inside. In this way, smoke complicated the older air pollution concerns of stale air and oxygen deprivation. Quite simply, "fresh" air could not be had during heavy smoke palls, either inside or out. Similarly, thick smoke blocked sunlight, long valued as a healthful force. By the early 1900s, physicians knew that sunlight could kill bacteria. Thus sunlight played an important role in preventing the spread of disease. Physicians began to associate thick smoke with the spread of tuberculosis, for example, since the smoke prevented sunlight from killing bacteria carried in dust.[37]

In general, however, as physicians began to express greater certainty about how smoke affected health, they tended to emphasize only the respiratory implications of dense smoke, particularly attributed to the inflammation and irritation of respiratory organs, and not the many ailments, from diarrhea to constipation to lethargy to criminal impulses, that nonprofessionals attributed to smoke. While new studies of the health effects of smoke provided stronger evidence of the reality of the connection, they also served to delimit smoke's evil. Smoke did not cause tuberculosis; it only made it worse. Smoke increased the incidence and severity of pneumonia and asthma, but not stomach ailments or the myriad other health problems nonprofessionals had attributed to impure air.

As researchers began to express greater certainty about the effects of smoke on health, the smoke abatement crusade gained new life. After the deep depression of the mid-1890s, which had interrupted early organized efforts to control coal smoke, dozens of cities developed sophisticated anti-smoke movements. In several cities reformers created highly influential single-issue interest groups. In even more locales the movement forced the passage of complex ordinances and the creation of powerful smoke inspection departments. Indeed, in an indication of the scope of the revamped crusade, cities as dissimilar as Cincinnati and New York supported very similar, influential smoke abatement movements.[38]

At the turn of the century, Cincinnati was a densely populated, dirty industrial city, long-reliant on bituminous coal. Trapped in a river valley flood plain, Cincinnati experienced frequent temperature inversions and concomitant smoke palls. The civic elite were painfully aware of their city's relative decline among the Midwest's industrial giants. New York, on the other hand, was the nation's largest city, proud of its fine residential avenues and its relatively clear skies, which it owed to its long reliance on the clean-burning anthracite of eastern Pennsylvania. Although an industrial city, New York lay open to sea breezes and was less susceptible to the temperature inversions that haunted the river city. Despite the differences in economy and environment, these two cities developed similar movements based on the same rhetoric concerning health and aesthetics. In both cities the movement intensified after 1905 with the creation of a single-issue antismoke league, after which both movements gained enough influence to set public policy. The results of the movements, however, were far from identical.[39]

The Woman's Club, an exclusive organization of 150 prominent women, initiated the antismoke movement in Cincinnati. Although businessmen's organizations had long discussed the issue of smoke in that city, and antismoke ordinances had been on the books since 1881, limited public activism had ensured limited success.[40] In 1904, this began to change when the Woman's Club's Julia Worthington, the wife of an attorney, sent a letter to Mayor Julius Fleischmann demanding that he enforce the city's ordinance. Worthington and other members of the club watched smokestacks and chimneys in the city and noted the thickness of the smoke they issued, in the hope that their observations would serve as evidence against smoke offenders in court.[41]

The following spring the Woman's Club invited Dr. Charles Reed, the noted Cincinnati surgeon and gynecologist, to deliver an address on the smoke problem. In what became an oft-cited and reprinted speech, Reed made clear that he understood women's particular interest in smoke abatement. Noting that women were martyrs to the growing smoke problem, Reed

3. The Volunteer Smoke Inspectors Might Help Some. As in other cities Chicago's middle-class women actively participated in the effort to control smoke. "Club women," the original caption read, "propose to help Inspector Bird spot the offending chimneys." Chicago *Record-Herald*, 19 April 1909.

charged that the rights of women "never seem to be considered by the manufacturer of that class who fancies that in the assumed interest of his business he has a right to manufacture smoke without let or hindrance. The extra drudgery in housekeeping imposed upon women," he continued, "is never taken into account by the company whose factories fill the air with soot that filters alike into the parlor and bed-room." Moving beyond the issue of cleanliness, Reed also emphasized the health aspects of the pollution, relating smoke to tuberculosis, catarrh, and other respiratory diseases.[42]

Later in the year, Dr. Julia Carpenter, the physician who had first led anti-smoke activism in Cincinnati in the 1880s, also addressed the club concerning smoke abatement. Following Carpenter's appearance, the women of the club's civics department drew up a plan of action. Club member Mettie Miller summarized the women's philosophy when she wrote, "[T]he doctrine of cleanliness must be urged in season and out of season, for without cleanliness there can be no health, no beauty." In addition to continuing their surveillance of individual chimneys, the women also made general observations of the city's atmosphere each day of the week to identify the largest contributors to the city's smoke pall. Noting that Sundays were much less smoky than weekdays, the women concluded that domestic fires, which were most active on Sundays as families prepared dinner, contributed only a small fraction of the total smoke cloud. Smoke, they determined, was largely an industrial problem. They also understood that the solution to the smoke problem would have to come through engineering improvements, and they identified several devices in use under the clean chimneys in the city, hoping that others might install effective equipment. To make certain that city residents understood smoke was unnecessary, the club made public its list of clean chimneys as a means of revealing the possibility of smokeless combustion.[43]

In 1906, Julia Carpenter, Julia Worthington, Charles Reed, and dozens of other concerned residents broadened their efforts to control the city's smoke when they formed the Smoke Abatement League. Over the next few years the league grew into a powerful force in the city, and with a membership list that resembled a "Who's Who in Cincinnati," it had more than enough political clout to influence public policy. Mrs. Charles P. Taft and her husband, the owner and editor of the Cincinnati *Times-Star*, became active members, as did Murray Seasongood, one of the city's prominent lawyers. Julius Fleischmann, the former mayor and one of the wealthiest men in the state, also joined the league, as did William Proctor, president of Proctor and Gamble, Seth Foster, president of Stearns and Foster, and William Alms, president of Alms and Doepke Company. Andreas Burkhardt, Mrs. Henry Pogue, and George McAlpin, whose names adorned some of the city's largest retail outlets, also joined. Even the president of the United States, Cincinnati native William Howard Taft, supported the league and paid annual dues. In total, over two hundred of the city's most influential residents joined the league, including many successful businessmen whose own firms faced the smoke question in their boiler rooms.[44]

The league did more than lobby city government, however, as they hired a superintendent, George Sealey, to conduct his own investigations and make citizen's arrests of smoke offenders. Within a year the mayor appointed Sealey to the office of Chief Smoke Inspector, and the league replaced him with Matthew Nelson. Nelson, an insurance salesman, would become the city's most important antismoke activist. His diligent work made the league the driving force behind smoke abatement in the city. The league kept the issue before the public, particularly via Charles Taft's *Times-Star*, and Nelson's activism led to considerable success. In 1910, the newly elected progressive Republican mayor recognized Nelson's hard work with an appointment as the city's Chief Smoke Inspector.[45]

Significantly, as in Cleveland, St. Louis, and Pittsburgh, the earliest activists in Cincinnati understood the need for a technical solution to the smoke problem, but they did not seek to force any particular solution upon businesses. As Reed made clear in a lecture on the activities of the Smoke Abatement League before a national organization of stationary engineers, "what goes on at the lower end of the stack is the exclusive business of three men, the proprietor, the engineer and the fireman. The League is not made up of engineers, nor does it employ engineers."[46] In other words, any citizen could identify offensive stacks simply by watching the color of the emissions, which is what the Woman's Club and the Smoke Abatement League had done, but only experts could offer the solution by examining the equipment. The league and the city allowed the proprietor and his employees to solve the problem in the boiler room. For the moment, reformers were content to focus on smokestacks and chimneys, which they considered the sources of smoke, not the furnaces below.[47]

Just three days after Reed delivered his influential 1905 speech in Cincinnati, the New York *Times* reported on the efforts of Charles T. Barney in New York City. Barney, a real estate agent and a resident of Park Avenue just three blocks south of Grand Central Station, hoped to spark an effective antismoke movement in New York through the organization of a smoke abatement league, just as Reed had done in Cincinnati. Barney began with a message to the mayor demanding more effective enforcement of the city's ordinance. Fearing that the abatement effort in his city had turned into a trade war between hard and soft coal interests, with anthracite dealers leading the charge against soft coal users, Barney hoped to redefine the smoke problem in the city while rekindling public support for pure air. As Barney noted, New York's ordinance did not outlaw soft coal, but only its smoke. He also complained that since the public outcry against the smoke palls that accompanied the anthracite strike of 1902, the Board of Health had gradually lost interest in smoke abatement, becoming inactive and ineffective.[48]

Under pressure from Barney and the organization he created, the Anti-smoke League, the Board of Health revived an old ordinance which required the use of a smoke consumer and stepped up enforcement. Aided by the Anti-smoke League and sixty police officers assigned to the Board of Health to investigate nuisances, the city launched dozens of cases against smoke offenders in the early months of 1906. Members of the league notified Health Department officials of offending stacks so that the city might initiate legal proceedings. More important, once cases reached trial the league located witnesses to the offending stacks who testified that the smoke caused them annoyance, as the ordinance required for conviction. League attorneys took very active roles in aiding city prosecutors in preparing cases for trial. As result, 132 of the 193 arrests for smoke violations in 1906 led to conviction, whereas in the previous two years, without league help, the city had failed to convict a single smoke offender. Obviously the assistance of the Anti-smoke League made a significant difference in the city's effectiveness, and apparently had some effect on the clarity of the air. One New York publication, *Medical Record*, indicated that the league and the city had made progress in the summer of 1906. "This method of frequent daily arrests," the magazine wrote, "has had its effect, and there are now few chimneys giving forth black smoke, except those of the electric power houses and the Rapid Transit Company and the Edison Company."[49]

The Anti-smoke League did take on all offenders, however; and even the largest companies in the city, including the Long Island Railroad, controlled by Pennsylvania Railroad, and New York Edison, the city's large electric company, received the league's attention. In both cases the league worked with the companies and the courts to negotiate an end to smoke production. The Long Island Railroad agreed to burn only coke and hard coal within the city limits, and New York Edison began a series of elaborate experiments using various smoke prevention devices. Supported by the league and emboldened by recent successes in city courts, the city's commissioner of health, Dr. Thomas Darlington, even sought the arrest of the president of the New York Central Railroad, William H. Newman, for his part in "permitting the company to burn soft coal on its passenger trains and at the roundhouse at One Hundred and Fiftieth Street." Police notified Newman of the charge and ordered him to appear in a Harlem court. In essence, the Anti-smoke League had gained enough power in the city, through its resurrection of the smoke ordinance, to force even the most influential of New York's polluters into arbitration or into court, as the league negotiated and sued for a cleaner environment.[50]

As in Cincinnati, those most active in New York's antismoke crusade defined the smoke problem in terms of health, cleanliness, aesthetics, and morality. Barney pledged to remain active until he could "find out definitely whether the clear blue skies that used to be one of the chief charms of a residence in New York City can be restored." For those less concerned about the appearance of the sky, the Anti-smoke League also claimed that smoke could cause serious health problems for city residents. When the New York *Times* applauded the Anti-smoke League, it declared that its work "in the interest of health and clear skies for the Greater City is to be commended and supported."[51]

New York's Anti-smoke League and Cincinnati's Smoke Abatement League both represented a mature social movement intent on exerting control over the urban environment, but the results of their activism differed greatly. New York's reformers met with much greater success, in that clean-burning anthracite still fueled much of the city, and activists could often convince the worst offenders to switch to smokeless fuel rather than face public scorn. In Cincinnati, however, circumstances prevented such a simple solution. With a greater differentiation in the prices of soft and hard coals, a switch in fuels to reduce smoke was economically impractical. Thus the Smoke Abatement League could force only gradual improvement in the city's air quality. Regardless of the individual outcomes of these reform movements, however, by 1907, in New York, Cincinnati, and dozens of other cities, middle-class women, physicians, businessmen, and engineers, acting through interest groups, had defined smoke as a problem which required immediate municipal attention. These reformers successfully argued that smoke posed serious threats to health, cleanliness, and aesthetics, and in the next decade every large American city would react to those threats.

It seems natural that antismoke activism would increase with the volume of smoke, but the relationship was not that simple. As the stories of New York and Cincinnati make clear, cities involved in the smoke abatement movement did not experience similar rates of economic growth, increases in coal consumption, or comparable amounts of smokiness. That some seventy-five American cities legislated against smoke within a span of twenty years indicates that much more than an increase in coal use and its smoke was at work. Although smoke did increase with the expansion of soft coal use, particularly in midwestern industrial cities, clearly no threshold of smokiness needed to be met for residents to identify smoke as a problem.[52] Actually, some cities developed antismoke movements as a means of preventing the nuisance, rather than as a reaction to an existing crisis. As a Philadelphia engineer, William Ingham, complained at the Franklin Institute's discussion of the smoke problem in 1897, "There is no smoke nuisance in Philadelphia. Our clear skies prove that." Ingham suggested that Philadelphia had taken up the issue only because other great cities had, including London, Pittsburgh, and Chicago, and that Philadelphia was only trying to keep current. Smoke abatement was in vogue.[53]

Clearly the national breadth of the smoke abatement movement was not simply a reaction to a national environmental crisis. Instead, the movement owed as much of its intensity to the growing reform fervor of the progressive era as it did to the growing smoke clouds. From the 1890s through the 1910s, urban environmental issues gained considerable public attention. As reform-minded residents of all classes and occupations looked upon their cities, they saw badly flawed creations. Some reformers blamed machine politics for urban problems and emphasized municipal reform, while others blamed immigrants and a degenerate working class, and emphasized temperance, slum clearance, and immigration restriction. Many blamed monopolistic trusts for stifling competition, and supported government efforts to regulate big business, while still others blamed the neglected urban environment itself, and worked to improve parks, playgrounds, street conditions, and the city's air. And it was not unusual for a progressive urbanite to hold all these beliefs and to support all these reforms. According to some reformers, urban society needed to improve its government, citizenry, business structure, and its environment.[54]

The question remains, however, why did progressive-era reformers care so deeply about smoke? Environmental reform was not just a rational response to a deteriorating environment. Rather, attitudes about the urban environment were changing as rapidly as the environment itself. These changing attitudes owed their origins largely to the wealth created by the industrial boom. Middle-class urbanites searched for more than economic security, as new wealth stimulated a search for cultural and environmental amenities. New parks, scenic boulevards, beautiful civic buildings, clean water, and clean air would allow urban residents to lead longer, more pleasant lives. Environmental improvement became an important part of the middle-class effort to build a better civilization, one more befitting an affluent society.[55] Nevertheless, many reformers began to demand a wholesome environment not simply as an amenity but as a necessary part of human development and happiness. Perhaps Charles Reed best summarized this new attitude in his message before the Woman's Club of Cincinnati in 1905, when he proclaimed that "to breathe pure air must be reckoned among man's inalienable rights. No man has any more right to contaminate the air we breathe than he has to defile the water we drink. No man has any more moral right to throw soot into our parlors than he has to dump ashes into our bed-rooms." These assertions Reed labeled "the ethics of the air," and we might recognize them as a mature environmental ethic. According to Reed, residents of industrial cities, no less than rural Americans, had a right to a clean, healthful, and attractive environment.[56]

What the women and men who spoke against the smoke nuisance had articulated in the late 1800s and early 1900s was nothing less than an environmentalist philosophy. Health, beauty, and cleanliness, these reformers believed, were as important to the progress of their cities and of their nation as were economic concerns. These reformers did not yearn for a preindustrial society, and they often expressed great appreciation for the advances in living standards wrought by the industrial economy. Still, while accepting that industrialization had made America modern, they argued that the health and morality of its citizens and the beauty and cleanliness of its cities would keep America civilized. The smoke abatement movement that gathered steam in the early years of the 1900s was an environmentalist movement, an effort to exert control over the urban industrial environment, an attempt to place beauty and health along with prosperity and profit as the ultimate goals of civilization.[57]

Ironically, the impetus for changing attitudes about the urban environment came as much from the achievements of industrial expansion as from its failures. The remarkably rapid economic success of the United States gave urgency to environmental reform. In the forty years since the Civil War, the United States had developed into the globe's preeminent economic power. Many observers noted that American civilization appeared to be approaching fruition. At the same time, in American cities, the centers of progress, the nation looked half-built and chaotic. If America was to fulfill what many Americans believed to be its destiny as the world's democratic model, then surely the cities would have to change. Chicago was no Rome, no Paris, not even London, though in a heavy pall of smoke it might pass for the last. In a reaction to the wide disparity between international economic success and local cultural and environmental failure, middle-class urbanites attempted to make their cities world-class, to make them worthy of the most advanced civilization the world had ever seen. For reformers, making wealth was not enough. The mark of civilization came in the spending.[58]

Surely no civilized city could hunker under a shroud of dense smoke. Although antismoke activists had immediate and tangible concerns, to many reformers smoke meant much more than discomfort, extra expense, and spoiled vistas. As Chicago's former chief health officer Andrew Young reasoned, "It is only by the strict regard of the rules of cleanliness and healthfulness that we are enabled to keep ourselves and surroundings in a condition befitting the common sense of enlightened humanity and civilization. Our atmosphere is no exception to the general rule." While productive factories were signs of modernity, factory smoke was "really the sign of barbarism." In Cincinnati, the *Times-Star* labeled the movement the "fight for the Anglo-Saxon Virtue of Cleanliness," and noted that little could be of more importance to an enlightened civilization than smoke abatement. Some reformers linked smoke abatement and other environmental reforms to the very survival of civilized America. Positive environmentalists, those reformers who linked environment to morality, worked to beautify and order what they saw as ugly and chaotic in an attempt to improve not just the aesthetics of their cities but the character of urban residents as well.[59]

With a sense of urgency and moral imperative, then, reform-minded urban residents formed organizations and committees, conducted studies and surveys, publicized their findings and their fears, and they all came to one conclusion: coal smoke was a grievous problem that required government action. Few suggested that the private sector be left to find a solution to the smoke nuisance without municipal involvement.[60] Industrialization had created an urban environment in which a cultural emphasis on individualism, autonomy, and private property rights had to evolve into a social philosophy based on organization, civic cooperation, and public rights. The industrial city required a municipal government which could effectively regulate the urban environment to protect the health of the citizenry and the beauty of the city. Reformers knew that municipal governments would have to play an instrumental role in the construction and protection of the modern American civilization. Although the antismoke movement would lead to greater government involvement in environmental regulation in the second decade of the century, success would continue to be elusive, and the smoke clouds would continue to thicken over the nation's industrial cities.[61]

The Atmosphere Will Be Regulated

Nevertheless, it may be affirmed that in the near future the use
of the atmosphere will be regulated with no less care in great
cities than is at present the traffic in the streets. This seems the
only possible solution of the atmospheric crisis with which civilization is threatened.   —*Current Literature* 43 (1907): 332.

Nuisance—In law, such a use of property or such a course of
conduct as, irrespective of actual trespass against others or of
malicious or actual criminal intent, transgresses the just restrictions upon use or conduct which the proximity of other persons
or property in civilized communities imposes upon what would
otherwise be rightful freedom.
          —*The Century Dictionary and Cyclopedia*, 1911

As women's clubs, businessmen's associations, engineering societies, and newly formed smoke abatement leagues pressured municipal governments to create effective antismoke legislation, city councils responded
with highly flawed ordinances. These new laws were brief and vague, and
when tested, courts often found them unreasonable and unconstitutional. In
the first decades of the twentieth century, however, municipalities crafted
more complex laws, which, when given the support of state legislation,
proved reasonable and constitutional, but only moderately effective. The
continued shortcomings of antismoke ordinances reflected the complexity of
the problem, including technical and economic issues not easily resolved
through legislation. City governments could easily outlaw smoke, but city
residents could not so easily find effective technical solutions, particularly at
economical prices. Thus two decades of municipal action resulted in much-
improved laws, but only moderately improved urban atmospheres.
    City officials used the municipal authority to regulate nuisances, derived
from English common law, to justify these new and often highly restrictive

smoke ordinances. Nuisance law required all citizens to use their property in ways that did not injure others. In nineteenth-century cities, most nuisances involved basic environmental problems: faulty or full privy vaults, garbage in streets, foul odors, rotting offal, and the like. In preindustrial cities, nuisance laws functioned retrospectively, since, for example, no one could complain about a foul odor from a slaughterhouse until they smelled it. As cities industrialized, however, municipalities expanded their authority under the nuisance doctrine. Cities began to pass preemptive laws against nuisances, forbidding the establishment of slaughterhouses in certain districts of the city, for example, to prevent foul odors from annoying residents. Through the nineteenth century, cities expanded nuisance law into an important means of regulating the urban environment, not just by establishing post hoc penalties but by shaping nuisance law into a tool for regulating development. Laws intended to reduce the smoke nuisance evolved along similar lines, first establishing fines for offenders, and later establishing a series of regulations for all businesses consuming coal as a means of preventing smoke production.[1]

In the late 1800s, the burden of enforcing nuisance laws fell upon health departments. Citizens complained directly to municipal health officials, who had the authority to require property owners to abate offending nuisances and to issue fines to those who refused compliance. This structure of nineteenth-century environmental regulation reflected the contemporary understanding that foul environments affected citizens' health primarily and that health officials could best determine what constituted a real environmental threat to the citizenry. Health officials responded most frequently to complaints of full privy vaults, bad drainage, filthy yards, standing water in cellars, and dirty alleys, all of which were thought to be sources of disease-carrying miasmas.[2]

For most municipalities, then, enforcement of specific ordinances against smoke fell within the purview of the health department, and city officials treated smoke like any other nuisance that threatened health. By 1897, Chicago, Cincinnati, Cleveland, Milwaukee, Minneapolis, and New York had all passed ordinances that empowered health officials to abate the smoke nuisance. Thus in many cities the earliest municipal efforts to control smoke came from officials who knew very little about smoke creation or the fireman's ability to control it. Health department officials generally had the same expertise as the women and physicians who led the effort to legislate against smoke; they knew offensive smoke when they saw it emanating from the stack. The conditions in the boiler room and the particular means of abatement did not concern them.[3]

Although municipalities assumed authority to regulate smoke emissions under nuisance law, no obvious means existed for distinguishing a chimney that emitted a nuisance from one that did not. Not all chimneys that emitted smoke did so all day long. In fact, few did. Some chimneys emitted dense smoke just once a day, at first firing in the morning; others emitted dense smoke intermittently the entire workday, as heavy smoke from the stack signaled the stoking of the fire. And, of course, stacks emitted all shades of smoke, some black, others various gradations of gray, and still others white. When considering the blackened sky, the dark streaks on white buildings, or soot stains on light shirts, the smoke problem seemed clear enough; but when attempting to formulate an ordinance that could abate the nuisance at its source, what appeared to be a black and white issue revealed a remarkable variety of shades of gray.

In the late 1800s, the initial antismoke ordinances of most cities avoided the complex realities of emissions and simply prohibited *all* smoke from bituminous coal fires, as they did in Pittsburgh and New York; or they outlawed an undefined "dense smoke," as in Milwaukee and St. Louis. Cincinnati's 1883 ordinance avoided the difficulty of defining which smoke created a nuisance by mandating that all furnaces in the city use "such efficient smoke preventatives as to produce the most perfect combustion of fuel or material from which smoke results." Not surprisingly, none of these approaches proved sufficient. City officials simply could not enforce to the letter those ordinances that outlawed all smoke, for as many businessmen pointed out, all fires emitted some smoke, if only for five or ten minutes a day when first ignited. These ordinances quickly faced legal challenges, and judges found them no less unreasonable than did prosecuted business owners. Ordinances that outlawed "dense" or "thick gray" smoke suffered the same problems as those that contained no adjectives at all. "Dense" remained undefined in these early ordinances, and proprietors could legitimately argue that the operation of their furnaces required the emission of some dark smoke. The few ordinances that required the use of some unspecified device, such as in Cincinnati, while more resilient in courts, proved no more capable of reducing smoke, as businesses needed only to show some effort toward abatement. Actual results were less important, and courts sympathized with owners who claimed the best available equipment simply did not abate the smoke. Judges often proved unwilling to fine firms that had already invested in new smoke abatement equipment, regardless of its performance.[4]

In 1893, St. Louis became one of the first cities in the nation to pass an effective antismoke ordinance. The city council actually passed two law simply outlawing the emission of "dense black or thick gray smoke" in open air, and another creating a three-man expert commission to study the smoke problem and permitting the appointment of inspectors. These inspectors had the authority to enter the premises of offending stacks. While inspecting polluting equipment, officials recommended improvements to owners and gathered information for the prosecution of recalcitrant offenders. The most active of the city's antismoke commissioners, engineer William Bryan, estimated that his office reduced the city's smoke by 75 percent in just four years under the new law, primarily by encouraging firms to install smoke abatement devices. Unlike antismoke officials in many other cities in the 1890s, Bryan had considerable expertise in steam engineering, and his knowledge allowed him to educate stationary engineers and business owners toward smokeless operation.[5]

Despite the progress made through enforcement of the ordinance, in 1895 several manufacturers challenged the constitutionality of the law, using a case against the Heitzeberg Packing and Provision Company as a means of appealing its validity. In November of 1897, after hearing arguments from the defense and from a lawyer representing the Buck Stove and Range Company and several other firms with smoke cases pending in a lower court, the Missouri State Supreme Court declared the ordinance void because it exceeded "the power of the city under its charter to declare and abate nuisances" and was "wholly unreasonable." As Supreme Court Justice James Gantt wrote in his decision, smoke was not a nuisance per se under common law, and the state had not passed any legislation declaring it such. That is to say, common law only treated smoke that caused harm as a nuisance.[6] Therefore, any ordinance passed by a Missouri municipality had to include some indication that the smoke emissions to be regulated caused annoyance or injury to some particular citizens or property. The city could not declare all dense smoke a nuisance and forbid it, which is what the 1893 ordinance did, but it could prohibit the emission of smoke that caused harm to person or property.[7]

With its 1893 legislation declared unconstitutional, St. Louis went nearly two years without a smoke ordinance or a smoke inspection department. In the spring of 1899, however, the city passed a new law which included a provision forbidding only smoke that caused damage, injury, annoyance, or detriment to any inhabitants or property in the city. As Bryan noted, the new legislation would require more evidence in cases against firms that emitted dense smoke, and it posed some serious practical problems. In a city full of smokestacks, how did one prove which stack produced the soot that damaged or annoyed particular people or property? In St. Louis, as in other cities, prosecutors began to rely on eyewitnesses from outside the smoke inspection department, particularly women who could testify that stacks near their homes emitted smoke that caused them annoyance, especially by soiling drying laundry.[8]

In St. Louis, then, the first legal challenge to antismoke legislation had interrupted municipal control of emissions, but after a two-year hiatus, new legislation left the city sufficiently empowered to regulate the city's air. Indeed, even as the city council passed the new ordinance in 1899, the St. Louis *Manufacturer* expressed concern that smoke abatement would greatly impede industrial development by placing an undue burden on firms that required cheap, dirty Illinois coal, the fuel that sustained most of the city's factories. Like many opponents of antismoke enforcement in other cities and at other times, the *Manufacturer* noted the importance of soft coal, not just to the industrial city's prosperity but to its very existence. During a published dialogue with Bryan which ran in the paper's editorial pages, the *Manufacturer* concluded, "Mr. Bryan will find the manufacturers ready to co-operate with any just measure, any reasonable plan to abate smoke, but will find the majority of them unalterably opposed to petty prosecutions, in some instances converted into persecutions by the penalties of small fines and continual annoyance." Obviously, some manufacturers understood the power that smoke abatement officers would hold under the new ordinance, and this voice for manufacturers in the city pleaded for lenience.[9]

The legal challenge to the antismoke ordinance in St. Louis in the 1890s was quite typical for the period. Similar appeals in other cities, including Cleveland and St. Paul, also contested the authority of local governments to regulate smoke emissions. In this first wave of legal challenges, then, opponents of smoke ordinances exploited the states' circumscription of municipal power in order to impede environmental regulation. The challengers in the Heitzeberg case, and in several similar cases around the nation, understood the severe limitations placed on city governments by state legislatures. They successfully argued that cities required specific state legislation before they could declare smoke a nuisance per se. In other words, the courts ruled that cities could not declare smoke a nuisance until state legislatures said they could. In Ohio, Cleveland, and Cincinnati officials lobbied the state legislature unsuccessfully for years before securing the passage of a sufficient enabling act. Indeed, the cities did not succeed until former Cleveland smoke inspector John Krause entered the legislature as a senator and coordinated a lobbying effort in Columbus. That the struggle proved so difficult suggested to some that powerful interests, including railroad and coal corporations, had waged a secret campaign to block the legislation in the state capital. Only with the assistance of persistent lobbying from smoke abatement societies and the Chamber of Commerce could Krause secure passage of the bill in 1911.[10]

As states passed such enabling legislation, opponents of antismoke regulation lost their most effective means of defeating municipal efforts to control emissions. Actually, the most influential opinion upholding the governmental right to regulate smoke emissions as a nuisance came from a District of Columbia appellate court which affirmed the power of Congress to regulate smoke in the capital district. The 1900 decision, *Moses v. United States,* denied a series of arguments made against the smoke ordinance passed by Congress in 1899. The plaintiffs, owners of a furniture store, argued that District officials could not prove that smoke from their stack had any negative effect on neighboring property or residents. They also argued that steps taken to abate the emissions from their building should have excused them from criminal prosecution. The court held both arguments irrelevant in upholding Congress's authority to declare smoke a nuisance, ruling that District officials need not supply proof of damage and that the firm's failed attempts to abate their smoke did not absolve them of the crime.[11]

The support of broad municipal regulatory powers in *Moses* found resonance in subsequent state cases, where the ruling in the District of Columbia provided precedent for the support of ordinances which declared smoke a nuisance per se in many other cities. Most importantly, *Moses* set the precedent followed in 1904, when a Missouri court affirmed that state's enabling act, which allowed cities of 100,000 persons to declare smoke a nuisance. *Moses* and the Missouri case, *State v. Tower,* then supplied the foundation for a series of rulings which not only affirmed the authority of states to empower municipalities to regulate smoke, but also maintained the broad powers wielded by cities in their attempts to improve air quality.[12]

Finally, in 1915 the United States Supreme Court heard one of the many appeals against municipal regulation of emissions. In *Northwestern Laundry v. Des Moines,* the high court rejected the argument that the constitution's fourteenth amendment protected private businesses from governmental intrusion via emissions regulation, which the appellant, a laundering firm in Des Moines, declared unreasonable and tyrannical. The court could not have been more clear in its rebuff: "So far as the Federal Constitution is concerned, we have no doubt the State may by itself or through authorized municipalities declare the emission of dense smoke in cities or populous neighborhoods a nuisance and subject to restraint as such; and that the harshness of such legislation, or its effect upon business interests, short of a merely arbitrary enactment, are not valid constitutional objections."[13]

Of course, wide court support for municipal regulation of emissions did not end the appeals by smoke offenders. The city of Chicago, for example, faced a legal challenge to its antismoke crusade in 1905, even though courts had repeatedly upheld the city's authority to abate smoke, a power the state awarded Chicago in an unusually broad enabling act passed in 1871. But after the city pressured the Glucose Sugar Refining Company, the huge glucose trust decided to battle the smoke law rather than control its emissions. As early as April 1904, Chicago's smoke inspector had identified the glucose company's factory at Taylor Street and the Chicago River as one of the city's worst smoke offenders. Several early prosecutions against the refiner ended in no fines, however, and in October 1904 an editorial in the popular daily newspaper, the *Record-Herald*, singled out the emissions from the glucose factory in its protest against the smoke problem. After noting the leniency of the courts in past cases, the editorial reported that "this company is now spreading its filth over all the lower West Side once more." Frustrated with the low fines and the continuing smoke, not just from the sugar factory but from many large polluters in the city, Chief Smoke Inspector John Schubert pledged daily arrests for consistent violators; and, emboldened by a new cooperative attitude from Justice Walter Gibbons, who presided over all smoke cases, Schubert promised more maximum fines of $100 per violation.[14]

The glucose factory burned 450 tons of coal every day, and as Schubert pointed out, by buying the poorest (cheapest) grades of coal the company saved hundreds of dollars a week. Schubert's goal, then, was to impose such heavy fines on the company that it would make no economic sense to use the cheap, dirty coal. Schubert received considerable aid from the *Record-Herald* in his renewed crusade, and from Dr. Theodore Sachs, a physician who announced that smoke was a principal cause of tuberculosis in Chicago, just as Schubert began his crusade. Speaking before the Council of Jewish Women, Sachs called for strict enforcement of the smoke laws, as the health of citizens demanded clean air. In an editorial published the same day as Sachs's remarks, the *Record-Herald* made certain that its readers would keep persistent offenders like the glucose company in mind as they thought about the health implications of smoke. "The day of excuses has long passed," the editorial concluded. "The need of action is imperative."[15]

As the pressure mounted and the fines added up, the refining company sought an injunction from a federal court to prevent the enforcement of the Chicago smoke ordinance. At the same time, the company began its own publicity campaign, touting the positive aspects of smoke, such as they were. The company's president asked, "What would Chicago be without smoke? It would be a way station." With dry humor, the *Record-Herald* extended this old argument, writing in an editorial: "What we need is more smoke. Our happiest imaginable state would be one in which the atmosphere rained soot to smear the city with its filth and keep the sun in perpetual eclipse." Anti-smoke activists sensed that the old conception of smoke as an indication of prosperity had lost much of its influence, allowing the editors to mock the Glucose Company's president.[16]

The *Record-Herald* went beyond the criticism of the Glucose Sugar Refining Company's Chicago factory and the weak defense of smoke offered by its president; it disclosed the company's poor record as an industrial citizen in another city. Not only did the glucose trust operate a smoky plant in Chicago but its New York City plant also produced dense smoke. The *Record-Herald* reported that several New York neighborhood organizations had joined forces and finances to protect "their homes and lungs from the company's smoke, soot and poison fumes." The newspaper concluded in no uncertain terms: "Our Glucose Sugar Refining Company needs to be literally beaten into submission to our smoke ordinances."[17]

Submission would have to wait, however, for after hearing the preliminary arguments, Federal District Court Judge Christian Kohlsaat issued a temporary restraining order, prohibiting the city from enforcing the smoke ordinance. Attorneys for the Glucose company argued that the ordinance was unreasonable and unjust, while claiming that smoke from its stacks did no one real harm, since the factory emitted smoke in a manufacturing district. After three months of hearings and contemplation, however, the judge ruled against the Glucose company, declaring the law valid and lifting the restraining order. After noting the city's clear mandate to abate nuisances, Kohlsaat concluded, "The welfare of the individual and the state, from every viewpoint, demands that public health, life and morals shall not be one whit compromised." Clearly pollution in pursuit of profits had suffered a setback in Chicago.[18]

In 1906, with the ordinance secure, Justice Gibbons expressed his disgust with repeat violators and exerted pressure on the Glucose Sugar Refining Company, renamed the Corn Products Company after Standard Oil swallowed the trust early in the year. In July, Gibbons issued a series of large fines to offenders, including several $100 fines, the maximum penalty. The glucose factory was far from the only concern fined heavily during this summer crusade, but it did accumulate several hundred dollars in penalties, as prosecutors repeatedly brought the recalcitrant offender before the judge. The company even took to its old argument that its smoke offended no one since the factory lay in a manufacturing district. Inspector Schubert retorted, however, that people did live near the factory, and, "they are poor people who need fresh air more than rich people."[19]

By October the glucose trust had had enough of Chicago's smoke law. Claiming that it did not have the capital to improve its equipment, the Standard Oil-controlled company decided to close its factory for sixty days to assess the situation and stem the accumulation of smoke fines. The company laid off several hundred employees while it considered its options. At the time of the closing, the city had thirteen smoke cases pending against the refinery. In late November the company settled the remaining cases and paid $400 in fines. By 1909 the Chicago plant lay empty and for sale, as the giant trust continued its reorganization, moving its production to a new state-of-the-art facility. Surely other considerations went into the decision to close the old plant, but the economic pressure of legal fees, court costs, and fines, as well as the substantial negative publicity, may help explain the timing of the sugar trust's abandonment of the smoky facility.[20]

The obvious effectiveness of the smoke ordinance when wielded by an energetic smoke inspector and supported by a sympathetic judge alerted Chicago's smoky industries to the real power the city could exert over coal consumers. An amended ordinance in 1906 brought new concern for the city's many railroads, which had been "practically immune" under the previous ordinance, and Inspector Schubert gave notice that city inspectors would henceforth pay considerable attention to locomotives. Schubert's enthusiasm in the effort to reduce railroad smoke would eventually ensure another challenge to the city's smoke law.[21]

Hoping to avoid conflict with the city, the Illinois Central took extra precautions, stationing private inspectors along its line to prevent smoke arrests. Still, during the first day under the new ordinance, city inspectors found four of its engines in violation of the law. As Schubert initiated this new tough campaign he bragged that with fines ranging from $10 to $100, he expected to add $10,000 to the city coffers through smoke prosecutions over the next year. "When we get fifty or a hundred suits against the railroads there certainly will be a strong movement started to abate the smoke nuisance," the *Record-Herald* quoted Schubert. Surely the railroads understood that they would bear the brunt of this revitalized assault on smoke.[22]

Schubert's attack on railroad smoke precipitated still more legal challenges to the law, but with the city's authority to regulate smoke recently upheld in court, railroads could only hope to cripple municipal regulation, not prevent it. After two weeks of prosecution under the new law, attorneys for the Illinois Central argued before Justice Gibbons that the company had done everything in its power to prevent smoke. Every locomotive carried the latest abatement equipment and high-quality coal, or so the railroad claimed, and if Illinois Central engines smoked, the blame lay only with the firemen and engineers who operated the offensive locomotives. Gibbons expressed satisfaction with the argument and concurred with the railroad, ruling that henceforth the employees operating smoky locomotives would bear the cost of fines, not the company. In an ingenious strategy, the railroad then supported its engineers in court and demanded jury trials, suspecting that no jury would fine a workingman for doing his job as best he could. Schubert smelled a skunk, and complained bitterly to the court. The attorneys for the Illinois Central asked for and received jury trials for the twenty smoke cases listed against their engines.[23]

The strategy of the Illinois Central, adopted by other railroads as well, meant that Chicago's smoke inspectors would now have to spend long hours in court, testifying before juries, and fewer hours watching stacks for violations. Schubert vowed not to give in to the railroads, however, and asked the city council for more men to compensate for the lost labor. The fight against railroad smoke had turned into a battle of attrition, and by early September, seven months after the new effort began, the city had prosecuted 231 cases against the railroads, and fines for these companies reached $3,100.[24]

The Illinois Central's experiment with juries proved short lived. After several jury trials ended in high fines, the railroad's lawyers complained to Judge Gibbons that they found it impossible to seat jurors who did not display some prejudice against smoke law violators.[25] The lawyers soon realized that judges, and not the people, offered the best protection against heavy fines. After the railroads recognized that jury trials only slowed the process and could not prevent continuing fines, the companies took a new tack. In an effort to weaken the smoke ordinance, the railroads, supported by the Brotherhood of Locomotive Engineers, proposed new amendments, including one which would have required inspectors to consider the number of rail cars pulled by smoking locomotives. Engineers, who had suffered dearly under the attack on railroad smoke, as companies issued fines and suspensions for employees found guilty of breaking the ordinance, spoke out in favor of the amendments. This new approach by the railroads also failed, however, as the amendments did not pass council, and the fines against the companies continued to mount. But the war on railroad smoke in Chicago had just begun, and in the coming years the companies would have to spend much more time and money to control their smoke or to limit the effectiveness of the anti-smoke law.[26]

The persistent public support for effective municipal and judicial action in the fight against smoke in Chicago mirrored changes occurring in many cities around the nation in the years surrounding 1910. The heightened activism of city smoke departments in Chicago, Cincinnati, New York, Milwaukee, Pittsburgh, and many other cities, revealed the significant changes in public conception of municipal regulation which had occurred over the previous decade. States had given cities the authority to regulate emissions, and many urban residents had come to assume municipal governments had the *obligation* to improve air quality.

The heightened demand for effective municipal regulation led not just to greater activism by city officials and to increased support from judges, but also to a new generation of ordinances passed in cities around the nation. The form of these new laws reflected numerous changes in municipal environmental regulation. Milwaukee's 1914 smoke control ordinance was typical. After four major rewrites and dozens of minor amendments since first taking effect in 1898, Milwaukee's ordinance had grown from four brief sections to twelve, some of which were quite lengthy. The 1914 ordinance stipulated that the chief smoke inspector be "by trade or profession a steam or mechanical engineer with an active practical experience of at least five years." The law prohibited the emission of dense smoke for longer than five minutes in any one hour and required that all new construction or reconstruction of steam plants, boilers, or furnaces receive a permit from the smoke inspector. Penalties ranged between ten and fifty dollars per offense. The law also conferred upon the chief smoke inspector the right to enter all premises emitting dense smoke. In essence, smoke served as a warrant for inspection and gave the smoke inspector the power to enter, without previous announcement, any boiler room or locomotive that issued dense smoke.[27]

The specifics of amended ordinances varied significantly from city to city, particularly in the number of minutes of smoke allowed per hour and in the sizes of fines for violations, but most newer ordinances contained the same basic components: the creation of a separate smoke inspection department, the requirement that chief inspectors be professional engineers, and the empowerment of smoke inspectors to access any offending equipment. In addition, many cities began to require businesses to obtain permits for all new construction and repairs of boilers and furnaces. The new authority wielded by antismoke officials reflected the shifting focus of the crusade. Cities want-

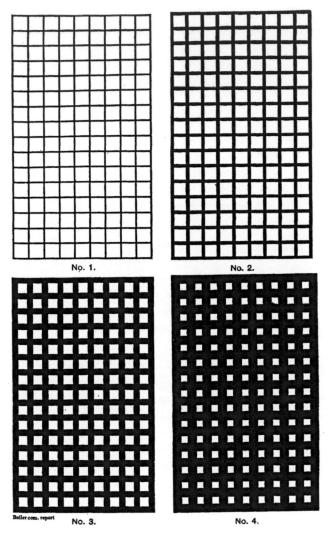

No. 1.   No. 2.

Boiler com. report   No. 3.   No. 4.

4. The Ringelmann Chart for grading the density of smoke. Held at arm's length next to smoke streams, the checked boxes of the scale created shades of gray that could be compared to emissions. Smoke the shade of numbers 1 and 2 generally was legal. American Society of Mechanical Engineers *Transactions* 21 (1900): 97.

ed more than officers to spot violations at the stack opening; they wanted experts who could examine boilers and furnaces to determine the cause of the smoke and to recommend a remedy.

Although most new laws did not offer a specific definition of "dense smoke," some ordinances did. In Pittsburgh, for example, late legislation indicated that smoke with a density of three on the Ringelmann Chart issued for longer than eight minutes constituted a violation. The Ringelmann Chart, used by smoke officials around the country, even where ordinances did not specify its use, allowed inspectors to stand at a distance from smoke and judge its density using a series of six checked boxes, each box a different shade, grading from white to black. The third box represented a density of 60 percent, indicating that smoke of that shade blocked 60 percent of the light passing through it.[28]

Not all smoke ordinances reflected the trend toward professional, scientific smoke abatement, however. New York's 1910 smoke ordinance left authority for smoke abatement with the health department, bucking the national trend toward the creation of separate smoke departments under the control of city engineers. The New York ordinance simply stated that no person "shall cause, suffer, or allow dense smoke to be discharged from any building, vessel, stationary or locomotive engine or motor vehicle, place or premise within the City of New York," but offered no indication as to how to judge the density of the smoke or as to how long a stack could issue dark smoke without fear of violating the ordinance. Although the law seemed archaic to some observers, after a year of intense activity the city made considerable progress against the nuisance, and the New York *Times* applauded the health department for its fine work. "One can stand on a housetop anywhere and see nothing but clean white smoke finding its way to the sky," one editorial claimed. The *Times* attributed the success to the great number of arrests for smoke violations over the previous year, by the newspaper's count more than two thousand in total.[29]

The city's new ordinance and the renewed vigor of health department enforcement meant considerable consternation for the New York Edison Company. Edison operated 114 boilers at its Waterside electric plant on the East River in midtown Manhattan, often producing dense smoke in the heart of the city, making the power company an obvious target for health officials. By July 1913, the city had issued nearly one hundred complaints against New York Edison in an effort to force the company to make preventative changes. Instead, Edison challenged the ordinance, and the city's Court of Special Sessions allowed the defendant's demurrer. In its decision concerning the smoke control section of the health code the court wrote, "The section is unreasonable and arbitrary because of its unqualified and sweeping character, condemning as a nuisance a thing that may or may not be a nuisance and because it makes no provision to cases where compliance is impossible." Using arguments that had worked in Cleveland in 1895 and St. Louis in 1896, Edison's lawyers had convinced the judges that employees could not operate the massive plant without creating dense smoke at least some of the time. The court believed that some smoke was necessary, and that therefore the strict enforcement of the ordinance violated the Constitution by depriving the company of property without due process of law. According to the court, the ordinance posed an unreasonable restriction upon the use of private property in the city.[30]

During the next week, the New York *Times* appealed to the city to challenge the ruling of the lower court. All coal, an editorial read, could be burned without smoke, and citizens had a great stake in the ability of the city to regulate such emissions. "It is uneconomic, it is ugly, and it is deleterious to health for all forms of life," concluded the *Times*. One column even offered scholarly aid, quoting a New York court of appeals case finding in favor of the city of Rochester's antismoke ordinance.[31]

Not all observers blamed the lower court for overturning the New York ordinance. *Power*, an engineering trade magazine that had long supported and publicized abatement efforts, roundly criticized the city's approach to smoke regulation. *Power* argued that the current legislation required judges and juries to determine which emissions constituted a nuisance and which did not. "With all due respect to the court judges," an editorial read, "they cannot be expected to know as much about the combustion of fuels as do combustion experts, and the interpretation of a smoke ordinance should be in the hands of those competent to weigh the enormity of a violation against the offender's ability to comply." In other words, *Power* argued that New York must create a smoke inspection department, run not by health officials but by engineers, experts who could determine the cause of violations and suggest practical remedies.[32]

Rather than rewrite its ordinance, however, the city appealed the ruling, and in December 1913 an appellate court dismissed the injunction against the ordinance. The court ruled the ordinance a proper exercise of the city's police power, arguing that the law need not be construed literally, that it should be assumed to prohibit only willfully discharged smoke which causes injury to citizens. Further, the appellate judges determined that in the city any dense smoke was a nuisance which warranted municipal regulation. "The marvelous growth of our manufacturing and transportation industries, and the rapidly increasing urban population, congesting in small areas, and the progress in scientific investigations with respect to conditions affecting health," the decision read, "have rendered it necessary in the interest of the public welfare to impose specific limitations and restraints upon the use of private property that were unknown to the common law." With *People v. New York Edison Co.* the court recognized the need to expand municipal authority, even through imprecisely worded ordinances when necessary, to protect the commonweal against the implications of private action.[33]

Like many of the nation's industrial cities, then, by the mid-1910s New York had overcome a series of legal challenges and temporary setbacks in its attempt to improve its air quality. The appellate decision against New York Edison reflected the influence of antismoke sentiment, even in a city most observers considered to be comparatively clean. The judicial support for New York's ordinance and the creation of sophisticated regulations in dozens of American cities signaled both a broadening and strengthening of the smoke abatement movement in the first and second decades of the century. From Boston to Salt Lake City, from Milwaukee to Atlanta, from the hard coal city of Philadelphia to the soft coal city of Birmingham, from industrial giants like Pittsburgh to transportation hubs like Altoona, by 1916 seventy-five American cities had passed antismoke ordinances. One wonders whether it had actually become a matter of civic pride to demand clean air even where the air was already rather clear. Undoubtedly some small cities took the opportunity to legislate against the big city problem of smoke as a means of announcing their own importance to the industrial nation. The Chamber of Commerce of Youngstown, Ohio, for example, invited Pittsburgh's Chief Smoke Inspector J. M. Searle to speak to their organization in 1913, as part of an initial effort toward creating a suitable ordinance in that city, even though some members expressed the fear that overregulation might restrict economic growth in the fledgling industrial town.[34]

In the same year, Baltimore, a large but relatively clean city, due to its sea breezes and ample anthracite supply, lurched toward effective smoke regulation. The Women's Civic Club initiated action when it organized a commission of prominent men and lobbied the mayor to support its study of the smoke problem. The commission in turn hired E. A. Thompson, a mechanical engineer, to "take up the question from an engineering point of view." Thompson acted as the city's chief inspector and with one assistant he investigated complaints and inspected power plants and new boilers. Thompson could claim some success in abating locomotive smoke, due to the cooperation of the Pennsylvania and the Baltimore and Ohio railroads, which had employed two inspectors each to watch the stacks of their own locomotives in the city, but he complained that the old idea that smoke means prosperity persisted in Baltimore and that as yet he did not have an effective law. Still, Baltimore had joined the growing list of cities actively engaged in controlling coal smoke.[35]

As the movement spread to additional cities around the nation, it also intensified in cities long engaged with the smoke issue, and in some cities the movement resulted in noticeably improved air quality. In Philadelphia a new ordinance took effect on 1 January 1905, initiating a new activism in that city's smoke department. While Chief Inspector John M. Lukens expressed appreciation for the cooperation he received from many of the offenders, Philadelphia too faced legal challenges to its antismoke law. But with court support, in both upholding the law and imposing fines on offenders, Lukens vigorously enforced the ordinance. By 1908 he could claim that the central city was "entirely free from the smoke nuisance," and that considerable progress had been made in the manufacturing districts. Lukens made over 1,500 inspections of stacks per year, but rarely initiated prosecutions, since most proprietors agreed to take steps to abate the smoke and avoid the fines. Some turned to automatic stokers, which gave furnaces an even supply of coal, eliminating the plumes that often accompanied hand stoking. Most plants, however, turned to hard coal, or a mix of hard and soft coal, to avoid prosecution. In general, Philadelphia polluters did not turn to new technologies to solve their smoke problem; they simply returned to the fuel that had kept that city clean for so many decades.[36]

Pittsburgh, too, experienced noticeable improvement in its filthy air in the years before World War I. After a court declared its antismoke ordinance null and void in 1911, Pittsburgh sought a new enabling act from the Pennsylvania legislature, which obliged in the same year. The city constructed a new ordinance, reconstituted its smoke abatement department, and reappointed J. M. Searle as chief inspector. Armed with the new legislation, Searle, and his 1914 replacement, J. W. Henderson, an engineer, began a campaign against smoke which led to a gradual but significant improvement in the city's air quality. The Smoke and Dust Abatement League, a private interest group organized in late 1912 at a special meeting of the Smoke Abatement Committee of the Chamber of Commerce, offered important support for Searle and Henderson. The league, composed of a dozen Pittsburgh organizations, including the Congress of Women's Clubs, the Engineers' Society of Western Pennsylvania, the University of Pittsburgh, and the Civic Club of Allegheny

County,[37] kept the issue before the public by organizing a series of exhibits.[38] By 1916, using data from the United States Weather Bureau, Henderson reported that Pittsburgh had reduced its smoke by 46 percent over the previous four years, even while the amount of coal consumed in the city continued to increase dramatically. In September of 1916, Vice-President Thomas Marshall remarked on the improving conditions while visiting from Washington. "Your smoke is disappearing, and yet you maintain your industrial supremacy," he said. "People of other cities can no longer tease you about coming from a city of dirt and smoke."[39]

Rochester, New York, also saw improvement in its air quality as the result of an effective antismoke movement. With the support of the Chamber of Commerce and local women's organizations, the city passed an antismoke ordinance in 1906. In what the city took to be an improvement on Chicago's respected statute, Rochester prohibited smoke lasting longer than five minutes every four hours, rather than the more typical six minutes every hour. After a slow start in enforcing the rigid law, progress became apparent, with some observers claiming that up to 75 percent of the city's smoke had been reduced after three years of enforcement.[40]

All estimates of improvement in air quality were just that—estimates. No one could supply solid scientific data to precisely quantify the reduction of smoke, but the trend was obvious. In several municipalities around the nation, antismoke ordinances, when given the support of state enabling laws, sympathetic judges, and influential allies in the business community and the press, did make noticeable improvements in the quality of urban air. At the very least, the movement had created a general consensus that municipalities could and should regulate emissions. Perhaps most important, however, the movement had largely convinced the nation that smoke was an unnecessary nuisance. As Harper's Weekly noted in 1907, no longer was a visibly active stack a "badge of great prosperity." Perceptions had changed, as many urbanites looked upon the dark clouds and saw no progress, only pollution.[41]

This changing public perception of smoke, even when combined with active enforcement of sophisticated legislation, proved incapable of solving the problem, however. The issue had many facets, and no single solution could satisfactorily control air pollution. Although people usually referred to the problem as if it were a single entity—"the smoke nuisance"—in actuality smoke came from a wide variety of sources for a wide variety of reasons, which significantly complicated the work of inspectors. The new generation of ordinances reflected a broad understanding of the need for expertise in abatement efforts, and the engineers who led most antismoke departments certainly appreciated the complexity of the problem that faced their cities. However, the new laws and the new expert staffs could not overcome certain technological limitations and economic realities.

While the tall stacks of industry and the centrally located chimneys of office buildings attracted considerable attention, they produced only a fraction of the smoke in most cities. In Chicago, for example, the Association of Commerce conducted an exhaustive study of the sources of smoke in that city and determined that steam locomotives created roughly 22 percent of the city's smoke in the early 1910s. Metallurgical furnaces contributed another 29 percent, while high-pressure steam plants, used in industry and offices, contributed 45 percent. The report also determined that domestic fires contributed only 4 percent of the city's smoke. Of course each city had its particular mix of smoke producers, and each city had to find appropriate legal and technological means to regulate emissions.[42]

Stationary boilers, those used in power plants, many industries, and in office buildings, provided the most fertile ground for reform because they contributed greatly to the smoke problem and, as they were stationary, inspectors could always find them and judge improvement. Engineers had also created several devices for use in stationary boilers, some of which worked quite effectively when installed and operated correctly. Perhaps most important for stationary boilers were automatic stokers, which fed fuel to fires slowly and evenly, thereby avoiding the large plumes that resulted from careless hand stoking. Other popular devices included steam jets, which fired hot air into fires to ensure complete combustion of the coal, and brick arches, which removed the fire box some distance from the boiler, thereby keeping the fire hotter and more likely to achieve complete combustion. Some businesses turned to down-draft furnaces, which were designed to increase efficiency and reduce smoke by pulling oxygen through the fire's coals. Unfortunately, each of these four types of smoke prevention devices had significant drawbacks. Automatic stokers, belying their name, required close and skilled attention to assure efficient operation. Steam jets tended to reduce fuel efficiency, and brick arches required more space and frequent repairs. For those looking to build new steam plants, down-draft furnaces seemed to reduce smoke, but a high initial cost limited their installation.[43]

Although most smoke prevention equipment fell within these four categories, a remarkable number of devices entered the market, and smokelessness became a major selling point, particularly in cities where smoke inspectors made arrests and courts issued fines. Unfortunately, not all salesmen were perfectly honest about their equipment, and often devices did not perform up to expectations, even after demonstrations promised good results. Not all alterations ended in failure, however, as a letter written to the smoke inspector of Cleveland in 1907 made clear. The author, W. R. Warner of the Warner and Swasey Company, wrote to the inspector in search of advice concerning his chimney: "I believe the general impression is that chimneys are used for the purpose of carrying away smoke, but this new one that we have been using for the past year does not seem to fill the bill in any sense, for it not only does not carry away smoke, but the bricks at the top even are nearly as clean and free from grime as when it was erected." Warner used this tongue-in-cheek complaint to make clear his point: smokelessness was entirely possible.[44]

Many engineers argued that smoke abatement often did not even require the installation of special abatement devices, claiming that architects could prevent smoke simply through proper attention to boiler room design. Architects often neglected boiler rooms in their designs, particularly for apartment and office buildings, where cramped equipment stashed in basement corners often did not have the room for proper operation. The boilers themselves, for example, often sat too close to their furnaces, thereby constantly cooling the air around the fire, causing excessive smoke. Engineers also argued that proprietors of manufacturing plants often ran boilers above capacity, stressing furnaces and reducing efficiency, and in the process creating dense smoke. If proprietors simply installed the proper equipment required for the work load, many argued, they could prevent much of the smoke from their stationary boilers.[45]

To combat inadequate designs, cities began regulating new boiler construction, and all new facilities required permits. In Milwaukee, the 1911 smoke ordinance instructed all persons and firms intending to construct or alter any steam plant, boiler, or furnace within the city to furnish a written statement concerning the specifications of the equipment and the stack, as well as a discussion of how dense smoke would be prevented. In Milwaukee smoke inspectors reviewed these building permits, but some cities created new offices for the work. In Chicago, the Department for the Inspection of Steam Boilers and Steam Plants issued permits only after a supervising engineer determined that the proper smoke preventing devices would be used. In 1906, Inspector Schubert claimed that this part of the ordinance had been quite effective, and that "new boiler plants installed under the provisions of this ordinance have caused but little complaint as to smoke." This may have been the most efficacious provision in most cities' smoke regulation, s

the booming economy of the early 1900s new facilities were constantly under construction. Through these clauses in smoke abatement ordinances,
cities had effectively expanded nuisance law into the regulation of industrial and commercial development.[46]

Proper design and installation of equipment in itself did not ensure
smokelessness, however, as great difficulty came from incompetent operation of furnaces, both new and old. Expert engineers claimed that, given
proper firing and reasonable load requirements, almost any well-construct-
ed furnace could operate without smoke. The key to smokelessness, these experts claimed, was proper handling by the firemen and the engineers. "Incompetent firemen," said John Krause, Cleveland's supervising engineer,
"are responsible for a great deal of our trouble." W. L. Goddard, a Chicago
engineer in the employ of International Harvester, noted, "The most expensive part of the steam producing element is almost invariably placed in
charge of the poorest paid man about the plant, the fireman." This hard and
dirty job attracted unskilled men, and the firemen who did excel generally
aspired to move up to better-paying, less physical engineer positions. Efforts
to hire skilled men to work furnaces generally ended in failure, as better jobs
or better pay lured away talented firemen. Many businesses did not even attempt to attract educated and skilled men to work and remain in boiler
rooms, for most plant managers continued to assume the positions required
only strong backs and continued to pay their firemen as if they were ex-
pendable. Smoke inspectors argued, however, that a good fireman could
save a company many times his salary in fuel costs simply through skilled
operation of the equipment. At the same time, he could also save the city from
the smoke nuisance.[47]

Firemen offered a different opinion. In the national union publication, the
Locomotive Firemen's Magazine, one writer expressed frustration with "college-
bred" officials who had burdened firemen with the smoke issue. "Nobody
loves coal smoke, and no one loves it less than the locomotive fireman," he
wrote, "but when a lot of 'cholly-boys' insist that these same poor fellows
have got to burn coal and not make smoke, they are piling up a whole lot of
grief for all concerned." Expressing pride in a craft which the writer argued
took two years to master, this fireman may have seen official interference in
the firebox as just another intrusion of management in the worker's sphere.
Firemen argued that only men experienced in their work knew how to properly control boilers, and that only improved furnace construction, not "any
new-born theory as to how a fireman should bend his back" would result in
decreased smoke.[48]

Locomotive firemen may have had special cause for complaint. While abatement in stationary plants progressed only slowly due to inconsistent performances from equipment, the expense of altering plants for smokeless firing, and the ineptitude of operators, the problem of locomotive smoke proved even more difficult to solve, and the penalties for failure often descended upon the workers themselves. The highly varied loads placed on locomotive engines meant equally varied requirements for pressure, which in turn required irregular feeding of locomotive fires. Unlike stationary plants, where fires were usually started just once and kept hot all day, thereby producing dense smoke only in the morning for a few minutes while the fire was cold, locomotives, particularly in urban switching yards and roundhouses, often had several fire starts each day, and each new fire emitted dense clouds of smoke. In the nation's railroad hubs, particularly Chicago and New York, locomotive smoke posed serious problems.

Railroad locomotives were not the only special cases, however. In Milwaukee, where the Milwaukee River divided the central city and provided an important avenue for the transportation of goods through the center of town, smoke from tugboats created a dilemma. While Inspector Charles Poethke claimed progress in combating smoke from stationary stacks by 1910, he admitted that tugboats still menaced Milwaukee with dense emissions. Tugs proved particularly offensive since the boats' stacks emitted smoke at nearly street level right in the center of town. The tugs acted as moving targets for Poethke, the sole inspector in 1909, who needed to keep offensive stacks in sight for six minutes to confirm a violation. In 1912, after the city had hired two assistants to aid Poethke in his inspections, tugs continued to pose a problem, so much so that Poethke requested that the city purchase a boat so that he might follow offending tugs on the three rivers that flowed through the city. Even without the boat, however, Poethke began to have greater success against tugs, as courts supported his prosecutions and additional inspectors allowed for greater coverage of the city.[49]

Steamships also posed special cases for smoke abatement, particularly in cities where jurisdiction concerns complicated enforcement efforts, including Detroit, Chicago, Cincinnati, and New York. Some steamship operators claimed immunity from municipal laws, noting that the Constitution gave the federal government jurisdiction over navigable interstate waterways. In 1913, Cincinnati attempted to secure jurisdiction over steamboats on the Ohio River, which separated Kentucky and Ohio, by lobbying Congress for new legislation. Antismoke activists enlisted Congressmen Stanley Bowdle and Alfred Allen, who represented the two Cincinnati districts, to take the issue of steamship smoke to Washington. The city assumed that the War Department retained jurisdiction over the river for such matters. However, an April federal court decision which allowed Detroit to regulate emissions from vessels on the Detroit River cleared the way for Cincinnati and other cities to begin rigorous enforcement of local ordinances against steamboats on interstate waterways, even without specific federal legislation.[50]

Smoke also posed jurisdictional problems on the waterways around New York City, but of a different sort, since very little of the smoke that hung over New York Harbor actually originated from vessels. Prevailing winds blew industrial emissions from New Jersey over the water. As early as 1899, the smoke had become such a nuisance that John C. Fremont, supervisor of the harbor, lobbied Congress for a law giving federal officials jurisdiction over smoke abatement in areas around the nation's harbors. Of particular concern for Fremont were the copper works on Constable Hook, whose emissions took considerable blame for diminishing visibility in shipping lanes in the Kill Van Kull, through which ships reached Newark Bay. The copper works even received blame for particular collisions in the harbor, one of which caused the death of a sailor and extensive damage to two vessels. Horatio McKay, commander on the Cunard steamer *Lucania*, complained that smoke had caused his ship to delay in the harbor, unable to pass through thick clouds. "The smoke nuisance in the harbor," he concluded, "is a perfect disgrace to such a city as New York."[51]

In 1900, New York Representative Nicholas Muller sponsored a bill in the House which would have empowered federal officials to take action to clear the air over New York shipping lanes. Of course the smoke from New Jersey affected more than just shipping, and when Muller's bill came up for debate in committee, representatives from the Staten Island Chamber of Commerce traveled to Washington to testify. Businesses on the island, particularly those reliant on fishing and oystering, lobbied for federal action. The Chamber of Commerce brought photographs of the smoke danger and petitions from concerned Staten Island residents. Unfortunately for these residents, and for Fremont and the sailors who passed through New York Harbor, the bill that passed out of the Committee on Interstate and Foreign Commerce never came up for a vote in the full House.[52]

The congressional setback in 1900 was one of many for proponents of federal air pollution regulation in the early twentieth century. Yet despite continued attempts by polluters to limit government regulatory power, by the second decade of the century many cities had passed sophisticated antismoke ordinances, formed professional smoke departments, and improved air quality noticeably. Still, progress in smoke abatement came slowly. Despite the optimistic reports of smoke inspectors around the nation, coal smoke continued to menace the residents of most of the nation's industrial cities. In some cities the smoke clouds actually thickened. Efficient regulation of the atmosphere posed a complex problem, indeed, as limited technological capabilities and the frequently high cost of abatement continued to impede rapid improvements in air quality.

Ironically, for the residents of Cincinnati only a natural disaster could bring relief from the smoke that clouded their city. In late March 1913, violent drenching storms set off deadly flooding in Indiana and Ohio. In Dayton and Columbus, rivers rose quickly, killing dozens of residents. Cincinnati fared better than its upstream neighbors. There the Ohio River rose slowly for days after the initial rains, slowly enough for residents to evacuate low-lying neighborhoods and for businesses to move their valuables out of shops and offices near the riverfront. Only one woman died in Cincinnati, when her canoe capsized as it crossed the Ohio. Still, the flood did not spare the city. For the second time in seven years the Ohio burst its banks and forced its way down city streets. The Central Union Railroad Station lay abandoned for days as the water nearly topped the rail cars in its yard. The water reached Second Street and, of course, rendered the public landings inoperable. High water forced manufacturing plants to close in both Cincinnati and Covington, Kentucky, just across the river.[53]

As the river receded over the next week, it dropped its heavy load of silt, leaving a thick layer of mud on once-submerged surfaces. Even as firemen used their hoses to force the muck back into the retreating river, at least one observer noted an ironic cleansing wrought by the dirty water. In a letter to the editor of the Cincinnati *Times-Star*, Louis T. More, vice president of the local Smoke Abatement League, wrote, "For the last few days locomotives and steamboats have not been running because of the flood, and it must have impressed everyone that the atmosphere has never been so clear or the sunshine so bright in this city." The waters that had so literally soiled the city had also rendered it smokeless. More continued, "It is startling to look at the Kentucky hills and see their outlines clear and distinct without any haze of smoke."[54]

More's letter to the editor emphasized the importance of smoke abatement to some Cincinnati residents. He did not point to the obvious irony, however, that the flood waters that lapped at the base of buildings on Third Street, closed the factories in lowlands on both sides of the river, prevented the docking of steamboats, and prohibited locomotive traffic through the city's center had accomplished in a week what progressive reformers had not been able to achieve in seven years of organized effort: the abatement of Cincinnati's smoke. Cincinnati, equipped with one of the nation's most exacting ordinances, served by a professionally staffed smoke inspection department, and supported by one of the nation's most influential smoke abatement leagues, had not yet found a permanent solution to its most serious pollution problem. Still, in the spring of 1913, Cincinnati residents could celebrate the respite that the heavy rains had given them, if they took the time to look up as they shoveled mud toward the river.

Priests of the New Epoch
Engineers and Efficiency

We are the priests of material development, of the work which
enables other men to enjoy the fruits of the great sources of power in Nature, and of the power of mind over matter. We are the
priests of the new epoch, without superstitions.
                    —George S. Morison, Civil Engineer, 1895

Expert—An experienced, skilful, or practiced person; one
skilled or thoroughly informed in any particular department of
knowledge or art.
                    —*The Century Dictionary and Cyclopedia*, 1911

If by the end of the first decade of the twentieth century the smoke
abatement movement had failed to diminish dramatically the coal smoke in
the nation's industrial cities, it had succeeded in forcing the passage of increasingly complex antismoke laws and in solidifying municipal authority
to regulate air pollution. It had also succeeded in spurring technological innovation among stationary and locomotive engineers. Indeed, as municipalities searched for solutions to their smoke problems, they turned to engineers to produce some tool or technique that could abate dark emissions.
Businesses, particularly railroads, hired engineers to reduce their smoke production; the federal government employed a staff of engineers to conduct research on the efficient combustion of coal; and local governments began to
hire engineers to staff their smoke inspection departments.

The lay reformers who had forced municipal action against smoke well
understood that although they could easily identify the problem, they could
not determine its solution. City Beautiful reformer Charles Mulford Robinson stated clearly, "[I]t is improbable that the reformer is fitted to give many
points to experienced firemen regarding the care of their boilers." Lay reformers understood that it required the expertise of engineers to solve the

problem they had articulated. Often they attempted to gain the information themselves, as had the Women's Health Protective Association in Pittsburgh in 1891 when they sought information concerning antismoke devices. More often, however, lay reformers simply encouraged the involvement of engineers who might apply their expertise directly. In at least one instance, lay reformers actually funded the research of experts, when the Women's Organization for the Abatement of Smoke in St. Louis requested and funded a Washington University study of soot in their city.[1]

Thousands of engineers took up the smoke issue, even without encouragement from lay reformers. Although the profession had confronted the smoke issue for decades, no consensus developed among engineers regarding the feasibility or even the desirability of abatement. Many engineers followed the logic of the lay reformers regarding the smoke problem, arguing against the filth, ugliness, and unhealthfulness of the smoke. Charles Benjamin, for example, spoke eloquently concerning the immorality of smoke before the American Society of Mechanical Engineers in 1905. Benjamin, who served as Cleveland's smoke inspector, announced, "I would see the time when the black smoke at the chimney top shall be an emblem of piracy as much as the black flag at the mast head, and when the question of profitableness of piracy shall be relegated to its proper place." For Benjamin, just as for most reformers, smoke was a moral problem, to be solved even at considerable expense, even at the expense of efficiency if necessary.[2] Other engineers interested in the antismoke movement were less certain of the importance of abatement. One member of Chicago's Western Society of Engineers, H. E. Horton, proclaimed in 1906, "We of Chicago hope to rival London as a great center of civilization. If London finds it possible to endure the smoke, we can undoubtedly do the same." Clearly for Horton smoke posed no real threat to American civilization, either moral or economic.[3]

Most engineers engaged in the smoke problem undoubtedly took a position somewhere between those of Benjamin and Horton, and—without developing complete consensus—smoke experts gradually came to agree on two important points in the second decade of the century: that significant smoke abatement was technically possible, and that the implementation of abatement practices and the installation of smoke control equipment could save coal consumers money over time. By the mid-1910s, general agreement on these two points within the engineering profession transformed the public's conception of the problem and redirected municipal control efforts. The public environmental problem articulated by lay reformers began to sound more like a private conservation issue.

In their work most engineers took a narrow view of the smoke problem, focusing on the efficiency of equipment and personnel. They attacked the problem at its source, the boiler room and locomotive engine, not at the chimney and stack, as had lay reformers. They moved the antismoke debate into the exclusive realm of experts, and engineers in the employ of municipalities sought a dialogue with the men working below offensive stacks, usually shunning prosecution of offenders in favor of education. By the end of the first decade of the century, lay reformers had largely succeeded in forcing the issue onto municipal agendas. As a reflection of their success, the national dialogue concerning smoke abatement became ever more technical. Although middle-class women and other lay reformers remained active in the crusade, they began to lose the ability to shape the discussion of the problem. As the relative importance of lay reformers diminished, their arguments concerning health, beauty, cleanliness, and morality gradually lost ground to engineering concerns. Antismoke engineers spoke mostly of efficiency, conservation, and economy. In the national, and at times international, smoke abatement dialogue, engineers much more frequently referred to smoke abatement as a sound business proposition rather than as a means toward creating a more healthful, beautiful, and moral civilization.[4]

The nation's growing reliance on engineers in the effort to control smoke was hardly unusual. Urbanites had come to rely on engineers to develop solutions to a wide variety of social problems, many of them environmental in nature.[5] Dr. William Goss, dean of the College of Engineering at the University of Illinois and director of an exhaustive study of railroad smoke in Chicago, concluded in 1915: "All progress in municipal development has been marked by a growing appreciation of science in its application to . . . public service." Engineers developed sewage, transportation, and water supply systems. They furnished electricity and natural gas, as well as the devices that consumed the energy. During the thirty years before the turn of the century, civil engineers had expanded their authority in government and in society. In an increasingly complex industrial world, engineers held the blueprints.[6]

Engineers boldly asserted their importance to the new industrial society, and as historian Edwin Layton argued, they struggled for the recognition and status they believed their position should command. Generally, engineers revealed a remarkable faith in the nation's industrial system, corporations, and the promise of reform through economic and technological advancement. Between 1880 and 1920 the number of engineers in the nation increased from 7,000 to 136,000. Through local clubs and national organizations engineers established themselves as an expert class. The American Society of Civil Engineers and the American Society of Mechanical Engineers, as well as other professional organizations, provided not just forums for the exchange of knowledge but bases for national influence as well.[7]

After 1910, the scientific management movement heightened the already substantial social prestige of engineers. In 1911, Frederick Taylor's *Principles of Scientific Management* became the guiding essay for the reorganization of American industry. Scientific management meant efficient management, which in turn meant an increasingly powerful voice for engineers in industry. Many engineers moved into management positions, where they made more than technical decisions, and where they based business decisions in technical knowledge. And, when the efficiency craze spilled out into the larger society, citizens clamored for more efficient governments and public services. Engineers again saw their stocks rise, as municipalities continued to depoliticize city management through the expansion of nonelective municipal engineering positions.[8]

The increasing influence of engineers in the smoke abatement movement, then, reflected the growing importance of engineers in the management of the nation, in both the private and public sectors. Engineers were hardly new to the smoke abatement effort in 1910, however. They had long participated in the debate, the movement, and in the search for a solution. In most cities, local engineering societies had actively participated in early discussions about smoke control, often supplying technical information at the request of non-engineering organizations. In 1892, the Engineer's Club of St. Louis supplied a detailed study of smoke for use by that city's reformers. In the same year, the Engineers Society of Western Pennsylvania also provided research and political support for the antismoke movement in Pittsburgh. In 1897, engineers gathered at the Franklin Institute to inform Philadelphia health officials about smoke prevention; their discussion, published in the *Journal of the Franklin Institute,* became an important source for antismoke activists around the nation.[9]

Using the trade press as an important tool for the dissemination of smoke abatement expertise, engineers gathered information from the United States and Europe. In 1903, for example, members of the Engineers' Society of Western New York developed an extensive bibliography concerning smoke abatement as part of a report delivered to their organization. The authors recommended that members interested in technical details refer to the dozens of articles they listed, largely drawn from *Scientific American, Engineering, Engineering Magazine, Engineering News, American Machinist, Cassier's,* and the

Journal of the Franklin Institute. Clearly, by the turn of the century engineers had created an international discourse concerning the practical means of abating smoke.[10]

Of course, the compilations completed by the Buffalo engineers rested on the practical research of other engineers and inventors. Since the inception of coal use, designers of stoves and furnaces had performed technical research to abate thick smoke. Indeed, no less a figure than Benjamin Franklin had endeavored to reduce smoke emissions from coal fires. In 1766, Franklin discussed plans for a stove that could burn coal without creating dirty smoke. Identifying smoke as unconsumed fuel, Franklin concluded that a stove producing no smoke would produce more heat. In the 1770s, while living in London, Franklin claimed to have finally built his smoke-consuming stove, with a downward draft, which he used for three years during his stay in England. By the late 1800s, some engineers referred to Franklin's work as the origin of smokeless design.[11]

As coal consumption increased dramatically during and after the Civil War, smokeless combustion gained greater attention among engineers and designers. Indeed, some designers dedicated their work to creating smokeless equipment. In 1860, Pittsburgh engineer D. H. Williams advertised his patented apparatus for smoke "combustion." Williams promised his steam jet required no structural alteration of the boiler, and that it would abate "entirely the smoke-nuisance and sparks," while saving 25 percent in soft coal bills. Six years later Samuel Kneeland wrote a thirty-eight-page pamphlet to advertise Amory's Improved Patent Furnace, which he promised offered both improved economy and smoke consumption. Kneeland's publication included dozens of testimonials from engineers in England and the United States who expressed satisfaction with the performance of the furnace, particularly in its great efficiency. In 1879, civil engineer D. G. Power wrote "A Treatise on Smoke: Its Formation and Prevention," a sixteen-page pamphlet dedicated to the selling of his new "smoke preventor." Even at this early date Power could note that the market contained numerous devices sold specifically for the reduction of smoke, although his treatise was quick to declare that no device, until his, could actually claim success.[12]

None of these early apparatuses solved the nation's growing smoke problem, nor did the dozens of other inventions that entered the markets in same decades. By the late 1880s, however, several devices had become so respected that engineers recommended them by name in trade publications.[13] The Roney stoker, manufactured by the Westinghouse Machine Company, and the Murphy "smokeless furnace," manufactured by Detroit's Murphy Iron

5. A few manufacturers, including the Murphy Iron Works and the American Stoker Company, used the smoke issue to help sell their products. Automatic stokers, like those here advertised, eventually became an effective means of reducing smoke for large fuel consumers. These ads appeared in the special "smoke issue" of the steel-trade magazine *Industrial World* in 1913 and 1914.

Works, both gained national renown for their smoke abatement capabilities. Actually, the Murphy Iron Works Company built its reputation in furnace equipment precisely because it could claim its furnace burned with no smoke. "Remove that big black pall of smoke hanging over your plant," one late Murphy advertisement urged.[14]

Manufacturers of furnaces and boilers were not the only firms developing smokeless equipment. Some of the nation's largest coal consumers also invested in smoke abatement research, particularly as antismoke activist offered radical solutions to the problem, such as the prohibition of soft coal or the electrification of steam railroads. For the Pennsylvania Railroad, for example, finding means to reduce smoke emissions could reduce fines, decrease operating costs by reducing the amount of coal consumed, and, perhaps most important, discourage the passage of municipal ordinances forcing railroads to electrify their lines in urban areas, which would require huge outlays of capital.

As early as 1894, efforts to control smoke through legislation in several cities sparked research at Pennsylvania Railroad's famed Altoona shops. In that year, motive power engineers conducted tests on steam jets and a ring device which blew steam into the base of the stack. The tests revealed that the ring device only succeeded in "whitewashing" the smoke with steam, not an insignificant result given the importance of color to smoke violations. The tests also found the steam jets unacceptable, as the device did not diminish smoke but did increase both equipment and fuel costs. Indeed, the Altoona engineers concluded that "anthracite coal offers, under existing conditions, the only satisfactory solution of the smoke question." Where anthracite was too expensive, such as in Pittsburgh, the engineers recommended only close supervision of the enginemen, not the introduction of new devices.[15]

Despite these discouraging early results, experimentation with antismoke devices continued at Altoona and other Pennsylvania shops. In 1910, engineers led by David Crawford, General Superintendent of Motive Power for the Western Lines, finished nine years of work on an automatic stoker for locomotives. In late 1910, Crawford reported to General Manager George Peck that the Columbus shops were applying about one stoker a week as part of an in-service test of the device. Some of the stoker-equipped locomotives were bound for Chicago, where the railroads faced an active antismoke movement demanding the electrification of steam lines. By the fall of 1913, 300 of the Pennsylvania's 3,430 steam locomotives carried the device. The small percentage of stoker-equipped locomotives nearly three years after its introduction reveals the company's unwillingness to bear large costs for unproven technology. New apparatuses of all kinds only gradually made their way into active locomotives as experience proved their worth. Just as important, the railroad installed antismoke devices only on locomotives that worked in or frequently traveled through cities with strict and enforced antismoke ordinances. Thus, even with the small number of stoker-equipped locomotives, the Pennsylvania system could produce results, as they hoped to in Chicago. Indeed, just a year after its introduction Pittsburgh's smoke inspector, J. M. Searle, called the Crawford Underfeed Stoker the "most important development in connection with the abatement of smoke in the past few years."[16]

Pennsylvania engineers continued their work on antismoke devices, searching for less expensive and more complete solutions. Just a year after the introduction of the automatic stoker, a committee of Chicago railroad officers confronting the antismoke movement requested that the Altoona shops conduct extensive research on steam jets and brick arches. Altoona engineers performed the tests in the summer of 1912 and issued their conclusions in the spring of 1913.[17] The final report, written by Crawford, concluded that steam jets could reduce smoke "to a very low amount" in "widely varied conditions," while simultaneously improving fuel economy. Crawford recommended the installation of steam jets in Chicago locomotives, a procedure which required little equipment or labor. Crawford did not recommend brick arches, which did not seem to reduce emissions and were expensive to install and maintain. An organization of railroad executives, the Chicago General Manager's Association, sent a copy of Crawford's report to each railroad in Chicago, and by midsummer of 1913 several railroads had already made alterations to their locomotives. The steam jets proved so successful that the General Manager's Association recommended that railroads equip all locomotives within the city.[18]

Led by companies hoping to sell smoke-abating equipment, such as Murphy Iron Works, and companies hoping to prevent the most serious implications of a successful antismoke movement, such as the Pennsylvania Railroad, private industry led in the early developments of more efficient, less smoky means of consuming coal. In fact, the private sector conducted the majority of research into the development of "smokeless" furnaces and smoke abatement equipment. However, even by the turn of the century, many engineers knowledgeable in smoke prevention argued that effective equipment already existed. As early as 1896, *Engineering News* noted that there were "on the market and in use to-day a number of complete combustion or smokeless furnaces." Smoke abatement awaited not the development of new equipment, the author claimed, but the installation and proper operation of existing devices. Four years later, the *Iron Age* argued, "The means of minimizing the smoke nuisance have been available for a quarter of a century." The *Iron Age* agreed that abatement awaited only proper operation of existing equipment.[19]

These statements confirmed what was obvious to nearly everyone. Just as the members of the Cincinnati Woman's Club quickly noticed as they studied the city's smoke in 1905, some stacks emitted very little smoke, while others emitted almost perpetual dark streams. The smokeless stacks gave evidence that coal could be burned without the production of offensive smoke. This did not mean, however, that every coal fire could burn cleanly. Since every furnace and boiler operated under unique conditions, concerning design, fuel quality, the magnitude and variance of the load, and the skill of the operators, smokeless operation actually posed significant problems. Even equipment designed for smokelessness could emit dense smoke under some conditions. In other words, no "cure all" for smoke existed.[20] Equipment that worked well with one grade of coal might not with another. Some equipment might greatly reduce smoke emissions when properly operated, but increase smoke when poorly operated. Some equipment worked quite well in controlled situations, when loads were even and fuel quality consistent, but in the variable real world some newly designed equipment offered little or no relief. The number and diversity of these devices created a confusing market for proprietors sincerely interested in reducing emissions. As Chicago Smoke Inspector Paul Bird warned, "The layman who is not an expert engineer, but merely a business man, is at the mercy of the salesman of smoke burners, preventers or devices, as the name may be."[21]

In the first decades of the twentieth century, the most important research concerning smoke abatement depended less on the development of new devices than on the systematic study of existing equipment. Engineers needed to examine extant equipment under varying conditions to note its situational effectiveness. Since businesses were wary of investing in unproven technology, this systematic research of devices went a long way toward convincing soft coal consumers that smokelessness was both possible and economical.

The Pennsylvania Railroad was a leader in testing equipment as well as in developing its own.[22] Inventors of smoke-preventing equipment contacted the Pennsylvania Railroad (and other companies) in the hope that the line would first test and then install their devices. The Pennsylvania's Test Department took these solicitations very seriously, giving blueprints to engineers for review and sending out representatives to meet with the peddlers of the most promising devices. Even the vaguest of leads could arouse significant attention from Pennsylvania personnel. In 1903, for example, mechanical engineer A. S. Vogt met with a Mr. Hughes, developer of the Hughes Smoke and Spark Preventer. Vogt came away unimpressed by Hughes's double brick arch design, concluding that the device would be expensive to maintain and a nuisance to boiler inspection. Although Vogt determined the design was hardly novel and most likely impractical, he concluded his report to Alfred Gibbs, General Superintendent of Motive Power, "Since, however, there is an effort to suppress smoke as much as possible, especially in and around Philadelphia, it may be worth while to take the matter up." Obviously, antismoke agitation along the Pennsylvania lines inspired the company to locate effective abatement equipment if it could not develop it on its own.[23]

This earnest desire to find a technical solution to the smoke problem led the Pennsylvania down many dead ends. In 1906, the company followed up on an ambiguous report of smoke prevention from M. J. Mulvaney, writing from Vandergrift, Pennsylvania. The Test Department sent engineer G. E. Rhoades to Vandergrift to investigate. Rhoades discovered that Mulvaney's solution was "a preparation to be sprinkled upon fuel to prevent smoke." Rhoades's report to E. D. Nelson, Engineer of Tests, revealed the humor of the situation:

As a demonstration of the effectiveness of his material, all that Mr. Mulvaney could show me was a small gas flame, at his home, in which he held a piece of wood, covered with the preparation. There appeared to be no evidence of combustion of either the wood or the preparation which dried out or baked on the wood and seemed to be entirely inert and to effectually prevent combustion rather than to promote it either with or without smoke, though Mr. Mulvaney assured me that it gave out a "beautiful heat."

Nevertheless, Rhodes noted all was not lost, in that he incidentally discovered that the compound was an effective insulator. He asked Mulvaney to send a sample to the Test Department for examination. "It might be interesting, if not very profitable," he concluded, "to have an analysis made when it comes."[24]

Railroad companies faced a particularly difficult problem. As conspicuous polluters they encountered public scorn and the possibility of political action which would affect them as a class of industry rather than as individual companies. Poor performances by one line might encourage the passage of costly restrictive laws for all railroads—laws which forced electrification or the use of anthracite coal, for example. Thus all companies running into cities in which smoke was an active issue endeavored to share information concerning smoke prevention devices and techniques. In 1909, for example, when a manufacturer of a locomotive device notified Alfred Gibbs that the New York Central was testing the equipment on an engine operating in New Jersey and New York, Gibbs requested that the Test Department send an engineer to observe. Within a month, engineer George Koch road the locomotive from Weehawken to Kingston and submitted a report on its performance. Koch noted that the equipment, which included a ring blower used to inject more air into a smoky fire, was not entirely smokeless. But since the coal was of low quality, Koch determined that the device was worth trying at the locomotive test plant in Altoona.[25]

This was familiar work for Koch, who in his years with the Test Department had traveled as far as Chicago to observe the operations of other lines. In 1907, Pittsburgh Smoke Inspector William Rea expressed interest in steps taken by the Chicago, Milwaukee, and St. Paul Railroad. Gibbs sent Koch with two "road foremen of engines" stationed in Pittsburgh out to Chicago to make a report, and, in the words of Superintendent of Motive Power D. M. Perine, "to show a desire to co-operate with Mr. Rea in the prevention of smoke." Koch's report, written with the road foremen, discussed the arrangements on the Chicago, Milwaukee, and St. Paul, and also the Illinois Central, Chicago and Northwestern, and Chicago, Burlington, and Quincy, which all entered the same Chicago Terminal. The Chicago, Milwaukee & St. Paul had no novel design, only the combination of brick arches and steam jets (also called smoke burners). The Northwestern was experimenting with a hollow brick arch, but largely depended upon careful firing and the stiff discipline of crews to assure compliance. The Illinois Central used steam jets with no brick arches, as the latter proved too expensive due to high maintenance costs. Koch and his co-writers concluded "that careful firing is the best method to prevent smoke and that any device was absolutely worthless unless the engineman and fireman co-operated in preventing smoke." In other words, the engineers in Chicago had been no more successful in rigging locomotives for smokeless operation than had the Pennsylvania engineers in the East.[26]

Whereas companies like the Pennsylvania Railroad did conduct systematic testing of abatement devices, most equipment went untested in the private sector, and salesmen continued to hawk equipment using testimonials as their only evidence of effectiveness.[27] In 1904, however, the United States

Geological Survey began its own investigations into coal consumption, justifying the research as a means of perfecting fuel use within the federal government, which consumed great quantities of coal on land and at sea. Using a special appropriation from Congress, the Geological Survey established a fuel testing plant at the Louisiana Purchase Exposition in St. Louis and began the systematic testing of equipment designed to reduce smoke and increase efficiency. The Survey's engineers gathered donated equipment from around the nation, shipped by companies eager to have their innovations approved by the federal government. Westinghouse sent a gas engine and an electric generator from Pittsburgh; Allis-Chalmers sent an engine from Chicago. Dozens of companies donated smaller equipment: gages, scales, belts, filters, crushers, etc. Although the Geological Survey focused primarily on the perfection of efficient coal consumption techniques, smoke abatement became an ancillary goal of its studies.[28]

In 1906, the Geological Survey published a lengthy progress report on its St. Louis coal testing. The most promising result of the tests involved the use of producer gas, derived from high-volatile coal and then burned to create energy. The Survey's tests showed that those coals with the lowest heat values, not coincidentally the smokiest coals, provided the best source of clean-burning producer gas. While the use of producer gas was not new in 1904, the tests did reveal the economic advantage of using bituminous coal in its manufacture. Indeed, low-grade coal, when transformed into producer gas and then burned in a gas engine, provided more than twice as much steam power than when burned directly under a boiler.[29]

While the results of the producer gas tests seemed promising in and of themselves, the Geological Survey's scope went well beyond this one means of deriving energy from coal. In 1907, Lester Breckenridge published "A Study of Four Hundred Steaming Tests," also conducted at St. Louis, and two years later Dwight Randall and H. W. Weeks published "The Smokeless Combustion of Coal in Boiler Plants," which summarized the results of tests conducted at St. Louis and at a second government facility in Norfolk, Virginia. These new reports made clear that smokeless consumption of coal did not require a massive shift to producer gas equipment. Rather, they concluded that all varieties of coal burned smokelessly under precise conditions. The reports advised that proper installation, proper settings, and a good match between equipment type and coal quality could effectively reduce smoke emissions. The government's tests also confirmed that smoky fires were inefficient fires. Even low-grade fuels, when burned without smoke, produced more heat.[30]

In 1909, the *American Review of Reviews* published an article entitled "Government Solves the Smoke Problem," in which the author, John Cochrane, optimistically and erroneously claimed that the Geological Survey had completed research that would soon make the nation's cities smokeless. Gas engines, central steam heating plants, and gas producer plants located near mines, Cochrane predicted, would soon replace less efficient and smokier means of energy production. While Cochrane grossly overestimated the ability of the St. Louis research to affect the nation's fuel consumption habits, he did correctly identify the federal government as a leader in smoke control research.[31]

In 1910, Congress created the Bureau of Mines, which assumed among its other responsibilities the Geological Survey's coal testing operations. The new bureau wasted no time in conducting and publicizing its work, led by Joseph Holmes, who had overseen government research in St. Louis from 1904 until 1907, and later at the new, permanent facility at Pittsburgh from 1908 to 1910. The Pittsburgh experiment station conducted a series of tests "for the purpose of studying the effect of different features of furnace construction and operation upon efficiency and smoke emission." The bureau published its results in various bulletins, but they also appeared in trade magazines, such as *Industrial World,* and in a somewhat less technical form in other publications, including *American City.*[32]

Although the federal government's research provided important information concerning smokeless combustion, the primary goal of its engineers remained efficiency, rather than smoke abatement. Still, government research made clear that smoke was not a necessary part of coal consumption. In other words, no longer could plant operators credibly claim that the use of available fuels simply necessitated the creation of dense smoke. Although a great variation in coal quality and power requirements prevented the identification of a universal solution to the smoke nuisance, research engineers did identify certain equipment as efficient and practically smokeless, at least when operated correctly.

Some of the engineers who conducted research with the Geological Survey and the Bureau of Mines continued their work on coal combustion while not in the employ of the federal government. After working at the St. Louis test plant, Lester Breckenridge conducted his research at the University of Illinois's Engineering Experiment Station, where he focused his work on Illinois bituminous coal. In 1907, he published an experiment station bulletin entitled "How to Burn Illinois Coal without Smoke." While this rather obscure venue for publication and the technical nature of the paper might suggest that Breckenridge's work had little effect on the smoke abatement movement, antismoke activists in Chicago and St. Louis were eager for any news concerning smokeless combustion of Illinois coal. Indeed, Breckenridge had already made a name for himself in Chicago, since he had delivered a series of lectures concerning the smokeless combustion of Illinois coal in 1904. The Chicago *Record-Herald* raved that this mechanical engineer was "probably the best equipped man in the United States to undertake a task of this character." Breckenridge's work received so much attention that the university reissued the bulletin in 1908, offering an additional ten thousand copies free to the public.[33]

For Chicagoans and others interested in the continued marketability of Illinois coal, few engineering tasks were more important than the development of some equipment or technique which could render the fuel smokeless, or at least less offensively smoky. For decades businessmen reliant on low-cost Illinois coal had argued that the very nature of the fuel prevented smokeless combustion. Breckenridge's work went far to discredit this argument and all similar complaints. He concluded: "[I]t seems safe to say that engineers now have sufficient information available to enable them to design boiler furnaces that will burn any coal without smoke." Breckenridge credited the systematic research conducted over the previous five years, particularly under the guidance of the Geological Survey, with eliminating the most powerful argument that smoke offenders had at their disposal: that no equipment existed that could burn cheap coal without creating smoke. After the wide publication of Breckenridge's work and that of other researchers, this argument gradually lost its currency.[34]

Another academic associated with the St. Louis experiment station, Washington University's Robert Fernald, also conducted research into smoke abatement outside his association with the Geological Survey. Indeed, the mechanical engineering department at Washington University became a leader in smoke abatement and coal combustion research under the direction of Fernald and other engineers. By 1915 the department had conducted a series of tests on coals available in St. Louis to determine efficiency and smokiness and had published its results. These tests and other research conducted at Washington University, including a study of soot fall in the city, provided valuable evidence for St. Louis's antismoke activists.[35]

Although the Washington University and University of Illinois research received considerable regional and at times even national attention, work at another academic institution, the University of Pittsburgh, provided the most important body of research into the smoke problem. In the fall of 1911,

Richard B. Mellon[36] provided funding for an extensive investigation of that city's smoke problem. Conducted through the university's Institute of Industrial Research, soon renamed the Mellon Institute, the investigation involved a thorough study of all aspects of the smoke issue, which researchers hoped would "reveal not only the nature, extent, and precise causes of the smoke nuisance, but also the remedies that would make its abolition possible and practicable." The institute gathered twenty-seven specialists to conduct research on various aspects of the smoke problem, and when complete, the staff consisted of seven physicians, five architects, four engineers, two chemists, an economist, psychologist, surgeon, bacteriologist, botanist, meteorologist, bibliographer, physicist, and an attorney.

Aside from publishing its results in nine volumes, from 1912 through 1914,[37] several members of the research team, particularly economist John O'Connor, a faculty member at the University of Pittsburgh, and chemist Raymond Benner, the chief fellow of the investigation, widely publicized the institute's results. Benner, who authored the bulletin concerning the effect of smoke on building materials, circulated reports on the smoke investigation in *Science, Iron Age, Coal Age, American Architect, Scientific American,* and the *Journal of Industrial and Engineering Chemistry.* O'Connor, whose bulletin on the economic cost of smoke became widely cited in both popular and trade publications, also served as the public relations director of the investigation and published in *American City, National Municipal Review, Popular Science Monthly,* and the *Pittsburgh Bulletin.* Through these many publications, reformers and city officials involved in the smoke abatement movement learned about the investigation and became partially aware of its major conclusions, even before the bulletins appeared.[38]

Upon hearing about the Pittsburgh study, many people interested in smoke abatement wrote the university in search of information. John O'Connor answered letters from a wide variety of parties, including Mrs. W. H. Snider, president of the Davenport Woman's Club, who wrote in July 1913, in search of an effective ordinance. O'Connor recommended Samuel Flagg's Bureau of Mines bulletin, which contained a number of sample ordinances. O'Connor also responded to dozens of letters from industries around the nation, including the George H. Smith Steel Casting Company of Milwaukee, whose chief chemist sought technical advice. The number and variety of the letters received by the Pittsburgh researchers revealed a deep interest in the scientific investigation of smoke. Many inquirers must have thought that finally the answers to this persistent problem would be found.[39]

Notwithstanding the series of articles and the numerous letters written by

Benner and O'Connor, the institute's bulletins provided the most important means of publicizing the investigation's findings. Six of its nine bulletins concerned the effects of smoke, one each for health, vegetation, weather, building materials, human psychology, and economic cost. The first volume simply provided an outline of the study, and the second provided an extensive bibliography of smoke-related publications in the United States and Europe. Thus only one of the nine bulletins concerned the causes of and remedies for smoke. As O'Connor noted a year before that volume's publication, however, "the most important branch of the entire inquiry is that of mechanical engineering." When it appeared in 1914, "Some Engineering Phases of Pittsburgh's Smoke Problem" represented the longest of the investigation's volumes. Although the study included a history of smoke and smoke abatement in Pittsburgh, the majority of its work relied on a survey of Pittsburgh plants designed to determine what businesses made smoke and why. The study's conclusion echoed those of Breckenridge and the several government publications that preceded it: "There is nothing impossible or wonderful about the smokeless combustion of even Pittsburgh coal, provided the proper methods are applied and the ordinary precautions taken."[40]

The Mellon study marked the height of smoke abatement research. Although subsequent projects added considerably to the understanding of smoke, its effects, causes, and remedies, the Mellon investigation set the standard for exhaustive scientific investigation. Together with research conducted by engineers in the private sector, within federal agencies, and at other universities, the Mellon study provided more than just data for reformers and municipal officials in search of a solution to the coal smoke problem.[41] It also provided scientific grounding for the antismoke movement, lending legitimacy to the reform effort as well as ammunition for the reformers.

Although engineers provided crucial information in the crusade against smoke, the influence of this expert class moved well beyond its capacity in research and education. As their authority within local governments grew, engineers gained substantial influence in setting public policy regarding emissions regulation. Antismoke ordinances revised after 1900 tended to require smoke inspectors to possess extensive engineering experience or training. By requiring that smoke inspection applicants pass civil service exams or possess exacting prerequisites, such as an engineering degree, reformers effectively removed political cronies from many important city positions. Of course by requiring that smoke inspectors have some engineering background, the new generation of ordinances did more than remove much of the political factor from appointments. Engineers appointed to head smoke departments often abandoned the prosecutory paths taken by previous inspectors, choosing instead to spend much more time in the boiler rooms below offending stacks than in courtrooms seeking penalties from offending proprietors.

By transforming the strategy of smoke inspection departments, engineers hoped to transform their effectiveness as well. As the Civic League of St. Louis summarized in 1906, "Practical results cannot be expected unless the department is filled with men who are qualified by training and experience, in the field of engineering." In other words, some observers had come to attribute the slow results from smoke abatement departments to a lack of expertise among inspectors. In St. Louis, for example, although an engineer had headed the city's first smoke department in the 1890s, in 1902 the city appointed Charles H. Jones, who was at the time the secretary to the chief of police, as chief smoke inspector. Over the next two years Jones had several deputies to assist him in his work, including a plumber, a machinist, and two clerks, none of whom, the Civic League complained, possessed the technical skills necessary to perform their duties adequately. Thus the lay reformers of the Civic League demanded that the city place control of smoke abatement within the hands of the engineering experts who could stop the smoke.[42]

In St. Louis and many other cities the solution to this problem was a clause in revamped antismoke ordinances requiring that smoke inspectors have engineering experience. By 1911, St. Louis had hired William Hoffman to head its department; Hoffman had sixteen years of experience as a mechanical engineer while working for the city's water department. Like other engineers placed in charge of smoke departments around the nation, Hoffman made cooperation the watchword of his office. "Cooperation," Hoffman said before his fellow members of the Engineers' Club of St. Louis, "is a valuable asset. Without it, the department is handicapped and progress is necessarily slow." Hoffman assumed that the primary goals of his department were to teach engineers and firemen how to abate smoke through proper technique and to instruct proprietors on the economic value of smoke abatement, even when abatement required the installation of new equipment. Cooperation, then, involved discussions with smoke offenders, a determination of the engineering problem, and the proposition of the most effective solution. Enforcement rarely involved prosecution.[43]

In some cities the replacement of non-engineer smoke inspectors was much less an effort to depoliticize the position through the expansion of civil service requirements than an attempt to shift power away from non-expert antismoke activists. In Chicago, for example, the efficient John Schubert, who had led very active antismoke campaigns in the summers of 1906 and 1907, was forced to resign when the city passed a new ordinance requiring that the chief smoke inspector have an engineering background. Schubert's ouster made way for Paul Bird, a steam expert from the Illinois Steel Company. The new ordinance also created an advisory board staffed by three mechanical engineers. For these positions city officials selected two consulting engineers, including Alburto Bement, an officer of the Illinois Coal Operators Association, and a steam engineer from the Corn Products Company, the very firm Schubert had most vigorously attacked the preceding year. Obviously the reorganization of the smoke department had greatly increased the influence of industry within the municipal regulatory body while greatly decreasing the influence of non-expert activists, who had found a powerful and active ally in Schubert.[44]

Although the changes in the Chicago department certainly shifted power within Chicago's antismoke movement, the institution of expertise may well have improved actual abatement. Both Bird, who left the smoke department to work for Commonwealth Edison in 1911, and his replacement, Osborn Monnett, created a smoke inspection department that received attention and praise from around the nation.[45] Monnett, a mechanical engineer who had served as an editor with two trade magazines and had worked for the Wheeling and Lake Erie Railway, developed a department geared primarily toward education and cooperation. Monnett and his deputies instructed firemen and engineers on how best to operate their equipment and offered advice to offending proprietors, essentially giving free and often extensive consultations to smoke violators. Monnett balked at court action against offending companies, and indeed upon his arrival he immediately showed his interest in cooperation by dropping pending suits filed by his predecessors.[46]

This new approach quickly gained favor in several cities, including Syracuse, New York. As a small industrial city, Syracuse did not suffer under the same level of pollution as St. Louis or Chicago, but local officials had actively engaged the issue since 1907. In that year, the Chamber of Commerce warned, "[I]t is not good policy to be too radical in the enforcement of the ordinance. The enforcement should be with discretion by officials possessing both technical and practical knowledge of smoke abatement." By 1914, the city had found such a man, Smoke Inspector Emil Pfleiderer. In correspondence with O'Connor during the Mellon investigation, Pfleiderer asked for any information that might make him a more effective "combustion engineer," volunteering that he was not engaged in "promiscuous smoke-abating by commands or threats of fines." Like other smoke abatement engineers around the country, including St. Louis's William Hoffman and Chicago's Monnett, Pfleiderer favored a go-slow, educational approach.[47]

As engineers came to dominate municipal efforts to reduce smoke, more than just the strategy of smoke departments changed. Engineers promoted their smoke abatement work as a means of improving efficiency, attempting to convince proprietors of smoky stacks that investments in new equipment in the pursuit of smokelessness would save money in the long run. In doing so, smoke departments dramatically increased the importance of a long-discussed but secondary argument against smoke emissions. Functioning as municipally funded consulting engineering firms, smoke departments made the promotion of fuel efficiency the focus of their abatement efforts. The protection of health, beauty, and cleanliness still garnered national attention, and the preservation of public health still formed the legal basis for government regulation of emissions, but for engineers within smoke departments these once-central issues could seem quite peripheral.

In 1906, several smoke abatement officials, including Milwaukee's Charles Poethke and Toronto's R. C. Harris, gathered in Detroit to form an organization through which they could consolidate their influence and professionalize their field. Meeting at the Wayne County Courthouse, representatives from Chicago, Philadelphia, Cincinnati, Cleveland, Rochester, Syracuse, Indianapolis, Denver, and, of course Detroit, created the International Association for the Prevention of Smoke, later known simply as the Smoke Prevention Association. The organization limited its membership to public officials dedicated to smoke suppression, but it did allow others interested in smoke abatement to be "associate members," meaning they could attend conventions and join in discussions but play no role in the association's decision-making process. The association set forth as its primary goal the creation of uniform smoke laws around the country. Defects in antismoke ordinances, members argued, were largely responsible for the slow progress toward abatement.[48]

In 1908, the growing association held its third annual meeting in Cleveland. The twenty-two members of the organization represented twenty municipalities, including most of the industrial cities of the Midwest. The next year, the association held its fourth annual meeting in Syracuse. Sixty-seven men attended, including municipal officials from twelve cities as well as representatives from the Geological Survey, including Samuel Flagg and Herbert Wilson. Most of the attendants came from the private sector, however, including several representatives from companies that manufactured "smokeless" equipment. The Jones Underfeed Stoker Company, for example, sent three men, no doubt in the hope that they might learn something about the smoke issue that would aid in the sale of their product and that they might impress upon city officials the effectiveness of their stokers. Attendants heard a series of speeches: Chicago's Chief Smoke Inspector Paul Bird discussed abatement efforts in his city; Dwight Randall discussed the relationship between smoke and the character of coals; Professor James B. Faulks of Syracuse University discussed the findings of his tests of boilers equipped with steam jets; and Wilson and Flagg, in separate presentations, discussed their work in the U.S.G.S.[49]

Although persons from disparate backgrounds brought disparate intentions to the annual meetings, and although those in attendance heard a wide variety of speeches, the primary goal of the association remained the establishment of standards for the field. Indeed, although the association began to allow nongovernmental officials to join in 1913, essentially the association continued to function as a professional organization for smoke abatement officials. These individuals, including Bird of Chicago, Searle of Pittsburgh, and Daniel Maloney of Newark, served as the leadership of the association, and guided the only long-term project of the organization: the standardization of the field. By 1915, at their meeting in Cincinnati, a committee listed ten general areas in which to establish standards, including ordinance language, office methods, stack observation, stack sizes, and the dimensions of certain boiler equipment. The committee also recommended the standardization of settings in eleven specific areas, including ratios of grate surface and size of retorts in furnaces, drafts in furnaces, the size and arrangement of brick arches, and the distance from fire to boiler surface. The committee, and the association, hoped to regiment the entire coal-consumption industry.

In an era when markets contained thousands of different furnace and boiler products and when many businesses still relied on archaic equipment, often used in tandem with the latest improvements, standardization would prove difficult, to say the least. Still, for the engineers who guided the association, it was the necessary first step in professionalizing their field. Only through standardization could experts collect practical data, set universal performance expectations, and document real improvement. Standardization was the first step toward putting smoke abatement on a firm scientific foundation.[50] The association had little power to enforce the standardization it so desired, but as a clearinghouse for smoke abatement information and a meeting place for those active in the movement, it did have significant influence over key city officials and over the press in the various cities where the association met, including Pittsburgh, Cleveland, Indianapolis, Grand Rapids, St. Louis, and Newark.[51]

Although quite sincere in its goal of abatement, the association actually proved to be a conservative force in the movement. Led by mechanical engineers whose knowledge and interests involved the burning of coal, the association concerned itself almost exclusively with the improvement of existing equipment and techniques to produce more energy and less smoke, and failed to investigate the promise of more efficient and cleaner fuel sources. Indeed, these men, so familiar with boiler rooms and furnaces, understood that existing equipment could operate smokelessly. They had seen it done and done it themselves, and they were more interested in spreading that knowledge than in discussing more radical and permanent changes, such as electrification or the extensive use of natural gas. As Inspector Viall of Chicago described to the association in Pittsburgh in 1913, often smokeless firing was just a turn of a wrench away, through the slowing of a chain grate, adjusting of a damper opening, or the straightening and sealing of breaching. Smoke abatement, these experts thought, would require no revolution in energy production.[52]

A review of the papers read before the organization reveals the engineers' goal of finding a coal-steam solution to the problem. Although many of the papers involved nontechnical issues, including talks on the law and publicity, members were much more likely to deliver (and hear) lectures on topics such as "Steam Jets and Their Uses," "Smokeless Locomotive Operation without Special Apparatus," or "The Development of Stokers and Their Relation to Smoke Abatement." When the engineers did discuss fuel, a common topic, they generally discussed the relative values of various coals and coke. Some papers reflected on the processing of coal into more promising fuels, either powdered coal or producer gas, but essentially all papers assumed that the smoke solution would involve coal and that bituminous coal would continue to play a central role in the nation's energy future. Not until the 1940s did discussion of oil become common, and natural gas, the cleanest burning fossil fuel, received essentially no attention during the first forty years of the association.[53]

The association's conservatism existed in part because of the dominance of mechanical engineers heavily invested in coal-burning technology, but also because of the absence of the broader movement's most radical group, middle-class women. In the decade before World War I, no woman delivered an address before the association or even registered as a full member. Most conventions included recreational activities for the women who accompanied their husbands, often taking them sightseeing as the men toured industrial plants to inspect operations. In 1907, for example, a special committee of women in Milwaukee entertained the wives of members by taking them to an art gallery and museum. The association, dedicated to technical solutions to the smoke problem, apparently had little need for the type of expertise women could muster, and the association embraced no radical voice inclined to question the go-slow, educational approach advocated by its engineering members.[54]

As cities reformulated ordinances to require their smoke inspectors to have engineering degrees or experience, the Smoke Prevention Association took on an ever more technical air. At the 1916 gathering, the association heard presentations concerning "Smoke from Low-Pressure Heating Plants," "Developments in Smoke Abatement Devices on Locomotives," and "The Development of Stokers and Their Relation to Smoke Abatement," and other technical speeches, all delivered by engineers. As the association grew in size and influence, its focus became more narrow and exclusive. By the late 1910s, as engineers discussed their research and methods, they also asserted their authority to find the solution to the nation's smoke problem.[55]

The growing importance of the Smoke Prevention Association and the increasing currency of engineering studies concerning smoke indicated the new dominance of engineers in the smoke abatement movement. By the late 1910s, engineers had assumed enough authority over the issue to alter the public's basic conception of the problem. The engineering emphasis on efficiency had partly displaced the discussion of smoke as a health and aesthetic problem, even outside engineering circles. Although many reformers still signaled the need to control smoke to preserve the health of urban residents, smoke-related economic issues gained greater salience. Even Kate McKnight, the president of the Women's Health Protective Association in the 1890s,[56] emphasized the cost of smoke to Pittsburgh's citizens during an important public hearing before the city council in late 1906. After repeating the common complaint from proprietors concerning the high initial investment required to install abatement equipment, McKnight reminded the council that the entire city bore enormous costs so that the owners might save a little by delaying capital improvements. The Pittsburgh *Post* summarized the theme of the council meeting with a front page headline: "Show It Pays to Stop Smoke." In Pittsburgh and around the nation, the economic aspect of the smoke issue had gained dominance. The new rallying cries for abatement hung on economy and efficiency rather than health and cleanliness.[57]

By the mid-1910s, the common perceptions of smoke had begun to change again. In the first decades of the movement, lay reformers had portrayed smoke as immoral and impure, the visible symbol of unhealthfulness and filth. Now, with the engineering solution dominating the discussion, activists were more likely to portray smoke as the exhalation of defective machinery, the symbol of failed mechanical operations, inefficiency and waste. The environmentalist arguments against smoke persisted, but conservationist arguments gained more attention. As C. H. Bromley, an editor of the trade magazine *Power*, could write to the New York *Times* in 1913, "The thing to remember is that the abatement of smoke is purely an engineering problem." In the earnest search for an engineering solution, the public environmental problem of smoke had largely been transformed, and the private economic issues of coal consumption and energy efficiency gained more and more attention. Surely the new dominant definition of the smoke problem indicated the continuing importance of smoke as an issue, but this new problem would prove equally difficult to solve.[58]

Smoke Means Waste

As to the economic side of the question: Smoke means waste of
coal; consequently, as fuel is the largest single item of railroad
operating costs, to suppress or even markedly reduce the
amount of smoke emitted from locomotives or power plants
cannot but have a material effect on these costs.
                                    —Altoona Railroad Club, 1910

Efficient—Producing outward effects; of a nature to produce a
result; active; causative.
                        —*The Century Dictionary and Cyclopedia*, 1911

In the years prior to World War I, the rhetoric of the smoke abatement movement shifted as cities restructured their smoke departments and
adjusted their practices. In their search for a scientific solution, engineers redefined the problem. Middle-class reformers led by women's organizations
had by the first decades of the 1900s slowly surrendered their position of authority on the matter, though they and other lay reformers continued their
efforts to control smoke in the nation's cities. As the movement became more
technical, it grew more exclusive, and the participation of non-experts became less important. The environmentalist movement now more closely resembled a conservation movement. Engineers understood that they were attempting to solve an economic problem, and as the economic argument
largely overshadowed health and aesthetic issues, concerns for property values and the economic cost of smoke also gained greater attention. Arguments
about citizens' natural rights to comfort and clean air gave way to issues surrounding property rights, particularly the right to adequate returns on investments, for both the consumers of coal and the victims of soot.[1]

Even as engineers gained authority over the smoke problem, public perceptions of smoke's health implications also changed, further diminishing the
potency of health-centered arguments. New research, culminating in the Mellon study released in 1914, tended to disassociate coal smoke from the nation's

6. Smoke Means Waste. By the early 1910s, engineers crusaded against smoke on grounds of efficiency. The engineering magazine *Power* here stated the argument succinctly. As the business owner looks over his coal bill, 15 to 30 percent waste goes out of the stack, and members of the community, including middle-class women, offer pointed comments. *Power* 39 (1914): cover.

most pressing health issue, tuberculosis. While many physicians and lay reformers had asserted in the late 1800s and early 1900s that smoke played an important role in the spread of the deadly disease, later studies largely disproved these allegations.[2] The first decades of the 1900s witnessed a major effort by city health officials, research physicians, and concerned interest groups to find a cure for tuberculosis. Although fresh, outdoor air remained an important part of the treatment for tubercular patients, the powerful forces searching for a cure put very little effort into smoke abatement. Instead, education concerning proper indoor ventilation, cleanliness, and disposal of sputum lay at the center of the antituberculosis movement in poorer neighborhoods, and removal of infected persons to suburban or rural sanitoriums dominated the effort to control the disease among the wealthier classes.[3]

Just as important, as the nation came to more fully integrate the germ theory of disease and scientific medical research into public conceptions of sickness, Victorian ideas linking health, cleanliness, and morality began to change. The population began to associate disease much more closely with specific microbes rather than with general environmental conditions, devaluing old conceptions of the relationships among cleanliness, physical health, and moral character which had been so important to earlier public health reformers. Influenced by racism, nineteenth-century conceptions of disease had associated the "immoral" behavior of poor immigrant families, the lack of cleanliness in tenement districts, and the high rates of death and disease. As scientific medicine disassociated disease from morality, the complex of ideas concerning public health also changed. And smoke, still an obvious cleanliness issue, and still a significant health concern, was no longer a factor in urban morality.[4]

In the first decade of the smoke abatement movement, the dominance of rhetoric concerning health, cleanliness, and aesthetics largely prevented businesses from effectively using economic arguments against control measures. Although those businesses resisting the movement had complained about the cost of redesigning plants for smokelessness and the extra cost of smokeless fuel, their most successful arguments against municipal action had involved the impossibility of smokelessness in coal consumption and the illegality of municipal ordinances. In the 1910s, however, with economic issues dominating smoke abatement rhetoric, businesses effectively used their own economic arguments against what they saw as draconian efforts by municipalities to control smoke.[5]

The fates of two very different smoke abatement movements, in Chicago and Birmingham, reveal the significance in the shift in the focus of the movement from environmental quality concerns toward economic efficiency issues. The profound differences between these two cities and the similarity of outcomes highlights the real problem facing so many American cities. Despite very different economies and political cultures, Birmingham and Chicago faced very similar air pollution problems, simply because both cities burned great quantities of bituminous coal. In the 1910s, Chicago was an industrial giant, one of the nation's largest and most vibrant cities and its principal railroad hub. It was also the home of one of the nation's oldest and most active smoke abatement movements. Birmingham was a much younger and smaller city. Heavy industries, particularly involving the production of iron and steel, dominated the city's economy, in a way that no industry could dominate a very diversified Chicago. Before the 1910s, Birmingham had no ordinance and no significant movement to force the passage of one. In these very different cities business interests used newly powerful economic arguments to stymie municipal action against smoke emissions.

Due to Chicago's long experience with heavy smoke and its nationally prominent smoke abatement crusade, that city deserves considerable attention. Despite its long-lived movement, in 1905 Chicago remained the home of one of the nation's worst air pollution problems. Perhaps second only to Pittsburgh in smokiness, Chicago dwelt under a thick pall issued from locomotives, a densely developed central business district, and an active industrial core. Although locomotive smoke troubled residents elsewhere, in no other city did railroad pollution have such a dramatic impact. By 1911, trains charged over two thousand miles of track within the city, with as many as 1,400 locomotives in operation at any one time. Not only the volume of traffic, but also the location of the rail lines made Chicago's locomotive smoke especially offensive. The Illinois Central drew particular attention, as its downtown terminal east of the loop brought smoke- and cinder-spewing locomotives through the heart of Chicago, through Grant Park, and between the city and its not-yet-prized lake shore. This Illinois Central line had retarded the increase of property values along the lake south of the loop and had prevented the city from gaining significant value from its lakeside downtown.[6]

Operators of stationary boilers already had several options available to help them avoid smoke: the installation of brick arches or mechanical stokers, for example, or the introduction of better coals, anthracite or semibituminous. But in locomotives space constraints delayed the introduction of an effective stoker, economic and safety concerns made the brick arch impractical, and anthracite did not perform well in many situations. In other words, before 1910 steam locomotives had made little progress toward smokeless operation, with careful firing combined with the use of semibituminous (Pocahontas) coal constituting the most reliable means of reducing the smoke nuisance. Given the centrality of the railroad industry to Chicago's economy and the manifest nuisance of locomotive smoke, debates concerning the abatement of railroad smoke eclipsed those of all other emission issues, and the electrification of railroads became the major focus of the city's movement for more than a decade.

As early as 1906, the Chicago *Record-Herald* began a campaign against locomotive smoke, concentrating on the Illinois Central. In an editorial, the newspaper argued that to date railroads had made a mockery of the smoke ordinance and predicted that only electrification could bring complete satisfaction. As the city council debated a resolution that would require electrification at the Illinois Central's terminal, the *Record-Herald* demanded action "for the sake of public health and comfort." For two years, however, neither the city nor the railroads took any action toward electrification, and the locomotive smoke problem only grew, but so too did the strength of the movement and its call for electrification.[7]

Although the intensity of the effort to control locomotive smoke posed a new challenge for the railroads, neither the idea of electrification nor the lobbying for its actualization were new. As early as 1897, before any steam railroad had opened a significant stretch of electrified track, the offensiveness of locomotive smoke had convinced some observers of the inevitability of electrification. "It seems that the day is near at hand," said Cloyd Marshall before the Chicago Electrical Association, "when the electric locomotive will find wide application within the limits of the great cities, and the steam locomotive will be outlawed." Marshall continued with a summary of the powerful economic argument in favor of electrification: "At present each railroad track is an unclean strip through the city, and the smoke, soot and dust make contiguous property undesirable. As a result the railroad routes are lined with huts, hovels and tumble-down warehouses."[8]

The railroads themselves expressed a great early interest in electric technology. Hoping to improve their service and decrease operational costs, steam railroads experimented with electric traction before the turn of the century. Even the later obstinate Illinois Central initiated its own investigation of electrification in 1892, in preparation for the World's Fair. Although the railroad's report on the subject concluded that electric traction technology remained in its infancy, and therefore did not warrant the capital investment required to electrify the lines to the fair, just five years later the railroad considered the new technology for the operation of its suburban lines. Despite this early research, the Illinois Central did not become a leader in electric experimentation, and the relatively smoke-free cities of the East, not filthy Chicago, first experienced the benefits of the electrification of steam lines.[9]

By the time the electrification debate heated up in Chicago in 1908, proponents of change could look to New York and Philadelphia for proof that the new technology was efficient and cost effective. Indeed, the experience of the two eastern cities gave tangible evidence of the benefits of electric traction. New electrified lines in Manhattan offered faster acceleration for stop-and-go suburban commuter lines, meaning electric trains could operate more efficiently and frequently. Champions of electric traction also argued that since electric locomotives could supply more power, trains could climb steep grades without losing speed, an obvious advantage for long freight lines which often had to limit the length of individual trains simply because such grades could throw steam locomotives off schedule.[10]

As proponents in Chicago emphasized, however, the most important advantage of electric traction was that it allowed railroads to operate smokeless locomotives. In the case of New York City, this quality had much more than atmospheric implications. Smokeless operation made travel safer in short tunnels and possible in long tunnels. In fact, the need for tunnel travel in New York spurred the most extensive electrification of steam lines in the nation, though not just because the railroads were eager to reap the benefits of electric technology. On 9 January 1902, a White Plains express train bound for New York Central's station slammed into another train which had stopped on the tracks in the tunnel under Park Avenue (then called Fourth Avenue). The engineer of the White Plains train had failed to notice three warning lights in the tunnel and by the time he saw the end of the parked train he could not bring his own to a stop. His engine barreled into a passenger car, telescoping its frame, and trapping sixty commuters. Firemen freed the passengers, but fifteen died in the accident. The engineer of the White Plains train, who escaped injury, stated in his defense that he had seen none of the warning signals due to the thick smoke and steam in the tunnel. The next day, as New York's attorney general began an inquiry, Mayor Seth Low announced his own conclusions. "The evident lesson of the disaster," he said, "is that electricity should be substituted for steam as the motive power in the tunnel. How it can be done is yet to be determined. I do not see how the accident could have been avoided under present circumstances."[11]

An editorial in the New York *Times* agreed with the mayor's assessment. Claiming that angry citizens should not blame the engineer, whom police had deemed sober and competent, the editorial concluded, "The responsibility rests higher up, and has rested there at least since 1891, when a similar accident was shown to have been caused by smoke and steam obscuring the signal lights." Seven passengers lost their lives in the previous accident, and railroad companies had since failed to follow instructions to improve ventilation in the tunnels.[12]

Government officials wasted no time in responding to the accident and the public outrage it created. Just a day after the accident a representative to the state legislature submitted a bill to force the railroads to electrify their tracks in Manhattan's tunnels. Although the bill did not pass in that session, a similar bill became law the next year. The state voted to provide $25 million for terminal improvements with the stipulation that no steam power could be used on Manhattan Island.[13]

The impossibility of operating a steam locomotive through tunnels of even greater length than those approaching Grand Central also convinced the Pennsylvania Railroad to electrify the new Penn Station in Manhattan and its proposed approach from Harrison, New Jersey, which would travel through a lengthy tunnel under the Hudson River. The company also electrified its newly acquired Long Island Railroad, which connected Penn Station with Queens and Brooklyn via a shorter tunnel under the East River. When completed, this electrification allowed the company for the first time to move passengers from New Jersey to Manhattan without a time-consuming and inconvenient transfer to a ferry to traverse the Hudson, and offered them an unbroken line across the nation's largest city, enabling passengers to move conveniently through New York and on to New England. By the end of 1910, the completion of Grand Central and Penn Station had dramatically improved access to midtown Manhattan, and electrification had resolved the issue of locomotive smoke for the entire borough, both above and below ground, greatly enhancing the property values around the island's new electrified stations.[14]

The Pennsylvania Railroad also undertook an extensive electrification of its busiest suburban lines in Philadelphia in the early 1910s. Increasing traffic had overwhelmed Broad Street Station, and given its prime downtown location, the railroad could not afford to expand the terminal. Instead, the company decided to electrify its suburban lines to Paoli and Chestnut Hill, increasing the capacity of the terminal without having to increase the size of its cramped switching yard. The electric trains simply moved in and out of the terminal faster, making room for more traffic. Electrification eliminated the need to load water and coal onto locomotives, reducing the length of stops. The electrified cars also required no turning, since they moved in both directions, unlike steam locomotives which required switching.[15]

Although the Pennsylvania Railroad took the lead in electrification of steam lines, its investments in electric traction in New York and Philadelphia did not signify a general support for the new technology and certainly did not indicate a desire to electrify all of its urban lines. In a 1909 essay on the smoke nuisance, Alfred Gibbs, the General Superintendent of Motive Power who oversaw Pennsylvania's shift to electric traction in New York and Philadelphia, noted that the cost of everything electric was "tremendous," and assured his readers that "the time has not yet come when the enormous outlay of capital for the purpose of electrification of the railways would be justified by the returns." Gibbs concluded that investments in other smoke-abating technology would prove more cost effective, including the development of better automatic stokers, then under way at the railroad's Altoona shop.[16]

Four years later, David Crawford, Pennsylvania's Superintendent of Motive Power on Lines West, echoed Gibbs's sentiments, and indeed borrowed his words, in a lecture before the International Association for the Prevention of Smoke. But he also expanded the argument against total electrification, even as the Philadelphia and New York investments were proving highly successful. Crawford made it clear that railroads still had an immense amount of capital invested in steam technology and, when combined with the equally large amount of capital required for the transformation to electric traction, the costs of general electrification were simply insurmountable. "As there are about 70,000 locomotives in the United States, representing an investment of about $1,400,000,000," Crawford concluded, "I am sure you will agree with me that some exceptionally favorable return must be apparent before they will all be discarded; especially so when their replacement involves an expenditure of many times their present value."[17]

Although the Grand Central and Penn Station electrifications proved immediate successes for New York and for the railroads, the precedent set in that city would pose public relations problems for railroads in several other important locations, including Chicago. Citizens began to ask why their cities did not deserve the finest, most advanced transportation technology. By late 1906, even before the completion of the great New York terminals, Americans had mounting evidence of the superiority of electrified lines, including the 1900 opening of Paris's beautiful Gare d'Orsay, Europe's first electrified station, and the early success of the New York, New Haven, and Hartford electrification in New York City. The booming interurban industry, with its electric trains running over small-gauged track, also gave ample evidence of the value of electricity to transportation. Indeed, by 1906 urban residents were well aware of the advantages of electric traction to consumers, particularly advantages related to its cleanliness.[18]

In 1906, with the construction of Washington's Union Station well under way, support for electrification of that terminal gathered momentum. The major line running north out of D.C., the Pennsylvania, had already invested millions in the terminal when the Senate took up a bill that would have effectively forced electrification of the station by subjecting locomotives to the city's strict antismoke law. Samuel Rea, third vice president, represented the company before a committee hearing and attempted to convince the senators that Washington had little in common with New York, where the rivers required tunnels and the tunnels required electrification. He also promised

7. Electrification of the Railway Terminals. The successful electrification of Grand Central Station gave electrification advocates in Chicago evidence of the many benefits of the new form of power. "When the railroads say that it would cost too much to electrify their Chicago terminals," observed a leading newspaper, "do they remember that electrification of the New York terminals made possible the reclamation of hundreds millions of dollars' worth of building property to the railroads in that city?" Chicago *Tribune*, 19 May 1913.

that the railroads would use semibituminous coal in the District and that smoke would not be objectionable.[19]

The Pennsylvania's policy to delay electrification in D.C., and other cities, was not made without the input of the railroad's engineers. Indeed, Chief Engineer of Electric Traction George Gibbs, who had been engaged in electrification projects for several years and who was heading work at Penn Station, specifically advised delay. Noting the developing character of electric traction technology and the high cost of being on the cutting edge, Gibbs advised the company not to decide in favor of electrification for "two or three years' time," and to wait another two years before beginning work.[20]

Rea clearly understood the need to delay action in Congress. In writing to the company's president, James McCrea, even before the Senate voted on the smoke bill, Rea noted, "What is occurring now in the National Capital will spread to every city." He argued that the railroads must take preventative steps, including changing the fuel used by locomotives within the city to coke or anthracite, and instructing enginemen and firemen in smokeless operation and giving them disincentives (fines) for their failures. "I think it is imperative that our Company at least should take steps in this direction," Rea continued, "not by simply talking about it all the time—because that we have done for many years—but give the people some ocular demonstration." Rea concluded, "Of course the only absolute cure for the smoke is to electrify, and it is towards that end that all these bodies are directing their attention, and against which we have exerted all our energies." Rea clearly understood the importance of electrification to smoke control, but he just as clearly understood the need for the company to control the pace of technological change.[21]

Less than a month after Rea's letter to McCrea, the Pennsylvania began to give the people "ocular demonstrations" of its desire to operate smokelessly (and with steam power) in the District. On the same day that the Washington *Evening Star* editorialized, "The railroad corporations are famous for promising reforms when reformatory legislation threatens," it also ran a story announcing the Pennsylvania's experiments with coke on a locomotive at its New Jersey Avenue yards. Less than two weeks later, in another happy coincidence for the railroads, the *Evening Star* announced that action in the Senate had held up the smoke bill on the same day it reported another successful coke test, this one at the Pennsylvania passenger terminal.[22]

The Pennsylvania did not act alone in its attempt to prevent forced electrification. All the major railroads in the District lobbied senators, particularly from the companies' home states. (Chesapeake and Ohio officials lobbied Virginia senators and Baltimore and Ohio officials lobbied Maryland senators, for example.) In the end, the lobbying proved successful, in that the delay in Congress outlasted the Senate's session. The railroads did not rest, however, nor did they go back to their smoky ways. Rea himself wrote the railroad executives urging them to set policies that would minimize smoke emissions, particularly through the regular use of coke. Rea wrote to George Stevens, president of the Chesapeake and Ohio, "[M]y own opinion is that the situation is too serious to jeopardize by using bituminous coal even of the highest grade for the reason that the moral effect of having coke on the tenders is more convincing than any exhibition of smokeless firing." Rea knew that the railroads were engaged in a public relations battle as surely as a battle against smoke.[23]

To keep railroad smoke out of the public eye the D.C. railroads formed a committee dedicated to the issue. This committee shared information concerning experiments with coke, including how to alleviate the buildup of clinkers when fueling with coke. The committee also organized the instruction of engine crews on smokeless firing and followed the cases of men accused of smoking. The railroads hired their own inspectors to watch the tracks throughout the District and to report the locomotives that issued black smoke. In October 1907, these inspectors watched over sixteen thousand engine movements and reported just ten cases of black and sixteen cases of dark grey smoke, a remarkable performance given the variety of situations, the diversity of equipment, and the number of employees involved. In the end, the railroads succeeded in their campaign, and electrification in the District awaited the Depression and Public Works Administration financing.[24]

In the Chicago debate over electrification, railroad officials, including those representing the Pennsylvania lines, made the same arguments Rea had articulated before the Senate committee. Chicago, they made clear, was no New York or Philadelphia. No special circumstances, like the need for subaqueous tunnels or for a superefficient terminal, made electrification particularly attractive to railroads operating in Chicago. Only the intense smoke problem kept the issue before the city and the railroads. Still, Chicagoans who favored electrification could not see the great differences between their city and the eastern cities. If it was practical to electrify on Manhattan, how could it not be practical to electrify the nation's railroad center? In 1908, Chicago Smoke Inspector Paul Bird, a trained engineer, sought to investigate that very question during a trip to New York. After meeting with W. J. Wilgus, the former vice president of the New York Central who had overseen the electrification of Grand Central Station, Bird concluded that the electrification of terminals in Chicago was "practical and feasible." He noted that elec trification of the New York Central had not only reduced traffic delays, but had also reduced operational costs, even when accountants included the capital charges for electrification. Following Bird's conclusions, Chicago's health commissioner, W. A. Evans, called for the electrification of Chicago's rails, claiming that engineers had proved its practicality. Noting that the city had spent $50 million on water purity, he wondered why it would not similarly invest in pure air. Even with the encouragement from Bird and Evans, however, the city council did nothing.[25]

By the late summer of 1908, the inaction of the city and the railroads, and the continuing smoke, inspired a vocal campaign by south side residents, led by Annie Sergel, president of Chicago's Anti-Smoke League, and other middle-class women. Using the same arguments made by women and other reformers around the nation for more than a decade, the women made a direct appeal to E. H. Harriman, who controlled the Illinois Central. The women also initiated a petition campaign in south side neighborhoods, hoping that organized activism might spur the city council in a way that individual complaints had not. The petition indicated that the undersigned found smoke from locomotives to be "inimical to the health, welfare and opportunity to live in ordinary comfort and decency." The petition also requested that the city require electrification of the Illinois Central.[26]

In response to the women's activism, the railroad initiated another study of electrification, its third in as many years. The company released a telegram announcing the decision made in New York: "It is the purpose of the company to go into this question, fully and immediately, securing the most able experts in the country to investigate the matter." The Anti-Smoke League remained unimpressed by the pledge for more investigation, and called for action, not contemplation. Sergel noted, "We are seeking something more than promises."[27]

The women of the Anti-Smoke League did not wait for the results of the new study. Meeting again at the home of Annie Sergel, the women passed a resolution which announced their intention to delay their housework so that they might more fully dedicate themselves to the cause. They also resolved to "spend no money in the acquirement of furniture or of objects of art with which to decorate our homes, nor of gowns to beautify our persons; that we even pack away all unnecessary ornaments which we now have in use, thus preserving them from the hands of the destroyer; all this with the hope of emerging next year in most beautiful butterfly fashion from the smoky chrysalis of soot and cinders to which a money-loving corporation might forever have doomed us." The women had assumed their moral high ground and with a sense of the dramatic had denounced the greedy soilers of civilization.[28]

As the women of the Anti-Smoke League circulated their petition, the *Record-Herald* kept the issue before the public, with nearly daily reports on the progress of the "smoke foes." In early October, Mayor Fred Busse met with representatives of the organization and pledged support for their cause. The city council also passed a resolution encouraging the state legislature to pass a law enabling the city to force commuter steam railroads, including the Illinois Central and six others, to electrify their lines within the city. One week later, Mayor Busse met with Illinois Central officials, including Louis Fritch, the Illinois Central's point man on the issue, and the company's general manager, Frank Harriman. Although Fritch had previously advised caution for his company and patience for reformers, Busse emerged from the meeting confident that electrification of the central terminal was not far distant.[29]

On October 19, two hundred women and men accompanied the Anti-Smoke League's petition to city council. With forty thousand signatures and the backing of forty-two other city organizations, including the Chicago Women's Club, the South Side Business Men's Club, and the Chicago Electrical Club, the Anti-Smoke League had gathered considerable political clout, and the council appeared ripe for action. As the activists made their case before the mayor, they still couched their argument in terms of health and aesthetics, and they spoke in tones of moral imperative. The Anti-Smoke League had embarked on an environmentalist movement akin to those waged by the Women's Health Protective League in Pittsburgh, the Smoke Abatement League in Cincinnati, and the Anti-Smoke League in New York City.[30]

While Chicago's railroads continued to delay action in the face of the league's pressure, locomotive firemen displayed more overt opposition to electrification and the movement generally. In December, John J. Hanahan, the grand master of the Brotherhood of Locomotive Firemen, visited Chicago from Peoria to protest against the smoke ordinance. Hanahan argued that the railroads simply fined firemen when the city fined the railroads, and that firemen, for their part, could not prevent all smoke. Hanahan, not inclined to take the railroads' side in the smoke war, indicated that company policies must change. At a meeting of the Brotherhood, which numbered forty thousand strong in Cook County, Hanahan called for an end to suspensions of firemen whose locomotives earned fines for smoking. Railroad officials in attendance of the meeting expressed sympathy, but offered no promise of change.[31]

The firemen had a great stake in the smoke debate. Not only did individual members face suspensions and fines from their employers, but electrification threatened their jobs. Electric traction would greatly reduce the need for men trained in the art of feeding coal into fires. As Brotherhood official Andrew Patrick Kelley argued, "Electrify the railroad terminals in Chicago and thousands of engineers, firemen and switchmen will be thrown out of employment." Although Kelley added, "I speak for the firemen, not for the railroads," clearly the railroads had a rare common interest with their employees, and a valuable ally in the effort to prevent forced electrification. Although the crusade threatened the livelihood of firemen, in both long and short terms, the firemen understood the need to reduce smoke, and they did not oppose technical changes out of hand. Actually, some new technology served to reduce the workload of firemen, without reducing the number of firemen necessary for operations. Automatic stokers, for example, reduced the burden on firemen in newer, larger locomotives that required remarkable amounts of fuel. Rather than feeding the fire directly, a multiple-step process, firemen operating an automatic stoker simply kept the stoker supplied with fuel and monitored its work.[32]

After hearing nothing from the railroad or the city on the progress of electrification, the members of the Anti-Smoke League expressed their concern in letters to the Illinois Central's Frank Harriman and Mayor Busse. Sergel, who authored both letters, wrote, "Never in our experience have we been more injured by the unhealthy dirt and fumes and annoyed by the puffing and screeching of the engines. The indignation expressed by the people through our petition has in no way abated." Responding through the press rather than directly to the league, Fritch retorted, "These women have no conception of the immense scope of the problem which confronts us. This problem involves the readjustment of our whole switch track system, which is a work for experts and cannot be concluded in a moment." In Fritch's eyes at least, the women had stepped out of their proper sphere when they attempted to force a specific solution to a problem only experts could fully understand.[33]

The activism of the league waned during the next five months, as the women allowed the Illinois Central to create its plan for electrification, and while Sergel took an extended vacation in Europe. Meanwhile, the state legislature failed to pass an ordinance enabling Chicago to force electrification. Feeling less pressure from the league and the city, the Illinois Central announced in October, one year after Sergel had organized the campaign against locomotive smoke, that it would not electrify its Chicago lines. In response, women's clubs around the city, including the Anti-Smoke League, vowed to renew their activism.[34]

As the women planned their next wave of activity, the city council again took up legislation which would impel railroads to electrify or find some other smokeless alternative. The two largest unions representing railroad workers sent representatives to speak before the Local Transportation Committee. P. J. Culkin, of the Brotherhood of Locomotive Engineers, emphasized the probable economic effect of electrification on railroad employees, noting that smoke should not be abated at "the sacrifice of the lives of the workmen." He also warned that passage of the ordinance would most likely force many men to move out of Chicago, as they searched for work on steam locomotives elsewhere. When asked if the use of coke would help, Culkin claimed that it would, and then he volunteered that better work from firemen would help as well. Railroad officials had already hired instructors and inspectors in a campaign to improve upon the performance of the workers, Culkin noted apparently without disdain, and he concluded that "they are obtaining grand results."[35]

After Culkin finished answering questions, Judge Waterman spoke on behalf of the Anti-Smoke League. Repeating the arguments made by the female members of the league for over a year, Waterman emphasized the aesthetic and health implications of smoke. He also compared Chicago with New York, calling the latter a leader in smoke abatement because of the electrification undertaken by the railroads there. Waterman obviously spoke with emotion: "Why are not the people of Chicago, the ladies in whose behalf I speak, the homes they live in, the women and children and the men and the boys of Chicago,—entitled to have their homes as pure and clean and the smoke nuisance driven away from their surroundings as much as the people of the great City of New York?"

Although Waterman was articulate, not everyone was moved. When Fireman Andrew Kelley spoke he began by complaining, "This looks to me like a woman's ordinance." Before making more derogatory remarks about the women of the Anti-Smoke League, Kelley noted the fallacy of the argument connecting smoke and health. "I have followed black smoke for fifteen long years, morning, noon and night, and look at the engineers and firemen who are here. They are specimens of the health and strength of the men who go to make up the railroad employees of this country." Thus Kelley held out his own fitness as evidence of the women's inability to judge the situation correctly, and he later questioned the propriety of women making judgements that would affect workingmen at all. Kelley, and other workers too, agreed with the railroads that when it came to solving the smoke problem, when it came to legislation, men, experienced men, ought to have the final say, and not women concerned with a little dirt.[36]

The last witness before the committee that day was Mrs. John B. Sherwood of the Anti-Smoke Committee of the Women's Club. She responded to Kelley's remarks with great sarcasm:

The gentleman of glib tongue was a fine specimen of a splendid workingman. He spoke of the women that ought to be at home. I want to say to him that I am going home in time to get my husband's dinner. There is quite time for that yet. The gentleman of such splendid health lives out of doors. We live in the house. Our homes are shut up. We get our husband's dinners. We take care of the children; we make the beds and we make the house as beautiful as we can and we work hard to make it clean. We do not get much fresh air. When we open the windows and let the fresh air into our homes the soft coal smoke comes in and we cannot breathe.

Sherwood's reaction to Kelley made clear that indeed her concerns were not about work, but about home. Smoke violated her home, just as it did the homes of the other active women.[37]

Railroad employees were not about to sacrifice jobs and job safety in return for cleaner homes for middle-class grumblers, however. Shortly after the Local Transportation Committee's hearing, two hundred railroad employees gathered in a show of opposition to the electrification bill. The men passed a resolution that read in part: "Smoke is at the present time an absolutely essential feature of industry, of work and the advancement of civilization. Before smoke is suppressed by law it must, in all justice, be proved that this can be done without loss and without injury to the work that smoke now accomplishes." The men also indicated that the clamor for electrification appeared suspicious, given that the supply of electricity in the city was controlled by a monopoly, and that coke offered an acceptable alternative to electrification.[38]

The men's concern about the potential control of railroads by the electric monopoly, Commonwealth Edison, was a red herring, since any large consumer of electricity, as surely the railroads would be, would likely find it economical to produce their own electricity at their own plants. William P. Rend, the Chicago coal dealer who years earlier had traveled to Pittsburgh to defend coal and coal smoke before the Engineers' Society of Western Pennsylvania, had sparked the monopoly concern in a lengthy letter to the editor of the *Record-Herald*. Rend, searching for any angle in his defense of coal and even coal smoke, flatly denied that smoke caused health problems and challenged anyone to prove that it did. He also played to public concern over the power of huge industrial trusts, a very significant issue in progressive politics. Rend suggested that Commonwealth Edison would be in a position to dictate terms to the city's industrial interests through its monopoly on power. "We are not quite ready yet to accept this vassalage and pay tribute to one of the most dangerous trusts ever formed in this country," he concluded. That the railroad workers picked up on this argument suggested that Rend's letter had some influence in the city.[39]

By early 1910 railroad employees representing several different unions, including passenger conductors, engineers, and firemen, had formed a committee to organize a lobbying effort against the proposed electrification, concentrating their efforts on city aldermen who faced upcoming elections. In 1911, railroad officials voted to take "whatever action may be necessary to work in conjunction (quietly) with a Committee of employees to oppose the passage of such an ordinance." The companies offered more than moral support for the workers' crusade, voting as a group to pay men their wages and $1.50 per day for expenses while they engaged in lobbying efforts.[40]

The pressure from the railroads and railroadmen, combined with state inaction, helped slow the movement toward electrification. In 1910 and 1911 the issue only sporadically reached the papers. However, railroads did continue a sincere effort to abate their smoke through some less expensive alternative to electric traction. In late 1909 and early 1910, the Illinois Central conducted very public tests with a new smoke abatement device. The railroad invited city aldermen to observe the trials in the hope that they would convince the council that abatement would not require electrification. The Illinois Central also began very public experiments with alternative fuels, namely coke and oil, on its urban lines. Neither fuel was new, even for use on locomotives, and the experiments probably centered less on technical matters than on economic factors. Both coke and oil were more expensive than the bituminous coal usually consumed by railroads, though much less expensive than electrification. Fritch indicated that the venture into alternative fuels did not mean an abandonment of its investigation of electric traction. Rather, he claimed that the railroad hoped either coke or oil might be used in freight trains on urban lines, which the company feared would be difficult to electrify.

The railroads also persisted with their argument that electrification would be economically disastrous and even potentially dangerous. Playing on the fears that electrified lines running through the city would pose a continuous threat of electrocution to workers, commuters, and pedestrians, one official from the Chicago, Burlington, and Quincy Railroad said, "It would break the biggest bank in the country to try to electrify local terminals and the city would be turned into a veritable slaughter-house if this was done. Men would be killed faster than in the civil war, there would be so many accidents." Although this extreme argument had little basis in fact, the Anti-Smoke League seemed to have lost its fight, and the city waited patiently under a cloud of smoke.[41]

If the railroads' strategy of delay through "ocular demonstrations" seemed similar to the approach taken in Washington, it was no coincidence. In the summer of 1912, the Pennsylvania's Samuel Rea, now vice president, wrote Second Vice President J. J. Turner concerning the smoke issue in Chicago, "I think what we were compelled to do in Washington, we ought to do in Chicago." More specifically, he called for the creation of a smoke committee with representatives from all the Chicago roads. Rea concluded, "I feel that if the railroads in advance take this situation in hand themselves, and appoint an independent Committee to inspect and look after this matter, both night as well as day, and publish reports of what they are accomplishing, it will largely satisfy all public demand and electrification may be put off for a ten year period." The battle to prevent forced electrification in Washington, then, had allowed the Pennsylvania to develop a successful strategy to prevent similar action in other cities, including Chicago.[42]

The Chicago railroads followed Rea's advice and the Washington example. Less than a month after Rea's letter arrived in Chicago, the General Managers' Association, representing all the city's roads, voted to create a subcommittee to consider the "advisability and practicability of establishing a joint bureau for smoke inspection."[43] By mid-December the joint bureau was up and running. This new bureau allowed smoke inspectors from any railroad to report smoking locomotives, regardless of their ownership, to the General Managers' Association. All major lines participated in the effort, though several companies apparently assigned road foremen to conduct the work of smoke inspector on top of their regular duties. A few of these men were not even stationed in Chicago. Despite the reluctance of some of the smaller lines to dedicate man-hours to the joint organization, or even to smoke abatement generally, the bureau did operate for several years. The bureau also functioned as a liaison to the city's Smoke Abatement Commission. In this way the bureau had both practical and political goals—to limit smoke from all Chicago locomotives and to show the city how much the railroads could accomplish without further regulation.[44]

At the height of the locomotive smoke issue in Chicago, the Chicago Association of Commerce, a public interest group consisting of many of the city's most important businessmen, undertook an extensive study of the smoke problem. On 29 October 1909, the association organized an eight-man committee, which included Smoke Inspector Bird, William Goss, an engineer at the University of Illinois, two other engineers, an economist from the University of Chicago, and three representatives from Chicago businesses. The committee investigated the engineering practicality and economic feasibility of electrification, and after eight months of study it presented its findings to the association. The committee bluntly stated that electrification was both practicable and feasible, and it recommended that plans for the switch to electric traction begin immediately, starting with suburban passenger lines.[45]

Although the report clearly reflected the committee's support for electrification, the Chicago Association of Commerce neither published nor publicized its findings. Instead, the association presented its conclusions to a number of railroad officials. Displeased with the findings, the officials argued that the report did not constitute a scientific study of the issue and demanded a new, more comprehensive investigation. With significant railroad influence, the association chose to ignore its own study and initiate a new, thorough analysis, to be funded by railroad money. In March of 1911, the association organized a new committee, with four representatives from the railroads, four from the city, and nine from the association at large. For the next four years the association's study would dominate the electrification debate.[46]

In December of 1911, the women of the Anti-Smoke League again lobbied city hall to support electrification. The mayor, however, responded that he would rest on the matter until the Chicago Association of Commerce had "finished its scientific investigation of the smoke question." Apparently, the women had lost much of their influence in the smoke debate. The mayor had chosen to wait for the findings of the male economic and engineering experts, whose advice would be forthcoming with the association's report, and to ignore the female voices that spoke for immediate action. Once welcomed into the mayor's office, the members of the Anti-Smoke League now had no particular standing in the electrification debate.[47]

For two years the Association of Commerce conducted its study, offering few results and no conclusions about the practicality of electrification. As the study dragged on and the railroads took no action toward permanent abatement of the smoke problem, alderman Theodore Long rekindled the issue by introducing a bill requiring electrification to the city's committee on railway terminals. Long's bill drew opposition from the railroads, as expected, but also from the Association of Commerce itself, which argued that action toward electrification ought to await the publication of the study, which was due in seven months, on 1 January 1914. The association went so far as to send experts to the committee to advise caution and delay. The city's prominent railroads, for their part, sent representatives to the committee, who intimated that they favored electrification, but that they opposed an ordinance that would force the switch prematurely. The railroads too argued that the council ought to wait until the publication of the association study, which had already spent two years and $250,000 in railroad funds collecting data.[48]

Long, angered by the association's official opposition to his bill and by the prospect of another year of inaction, made a lengthy, well-publicized speech before the committee denouncing the association and its relationship to the railroads. For the first time, Chicago citizens heard about the 1910 association study, which, in Long's interpretation, had been quashed by the railroads because it had arrived at the wrong conclusion. In confronting the railroads and their ally, the association, Long focused on economic considerations. While the women of the Anti-Smoke League had concentrated on health and aesthetic issues, Long merely mentioned these problems, and largely considered the economic benefits that would accrue from electrification: "It is estimated by competent engineers and real estate men that the valuable space that would be reclaimed by the abolition of the smoke nuisance would more than offset the total cost of electrification." Long even mustered evidence from his own experts, including W. J. Wilgus, the man who had engineered the New York Central's electrification. Long reported that electrification had actually reduced operating costs by up to 27 percent.[49]

Swayed by Long's powerful speech and his revelations concerning the duplicity of the association regarding its first study, the committee recommended the bill to the larger council. Long had succeeded in rekindling the locomotive smoke issue, but the issue itself had been transformed. Snubbed by the mayor and overshadowed by the association's "scientific" study, the middle-class women of Anti-Smoke League had lost their voice on the issue. The importance of health, cleanliness, and the beauty of the lake shore had dwindled, and now economic issues dominated the discussion. From this point forward, the city would hear only from experts, men trained in engineering and economics, on the issue of forced electrification. The city council did not pass Long's bill, stalling action toward electrification again, and with the transformation in the public debate, the moral imperative for change had been lost.[50]

In November of 1915, four years after its initiation and nearly two years after its expected completion, the association's smoke inquiry was finally complete. Not surprisingly, the study found that electrification to reduce smoke was neither advisable nor practicable. The study concluded that while electrification was technically possible, it was economically infeasible. When the report finally appeared, it was impressive, at least in magnitude: over a thousand oversized pages of text, charts, maps, graphs, and figures. It included a study of smoke, its chemistry and content; a study of emissions, indicating which sources produced the most smoke (stationary boilers) and which the least (steamships); a study of non-coal sources of air pollution, particularly dust; a review of the literature on the health effects of smoke, which concluded that physicians knew too little; and a lengthy study of the technical feasibility and cost of electrification.[51]

The association's report actually muddled the electrification matter. By taking the largest view of the smoke problem, the study had minimized the problem of locomotive smoke, arguing that it contributed only 22 percent of the city's smoke cloud and an even smaller percentage of air pollution generally when atmospheric dust was included. These findings directly contradicted the assertions of Smoke Inspector Bird, who had claimed just four years earlier that railroads created 43 percent of the city's smoke. The report also emphasized the importance of gaseous emissions, further diminishing the value of casual observation of stacks, an activity in which women had long participated.[52] Essentially the report concluded that the non-expert reformers, the women who had begun the electrification debate eight years earlier, were wrong. Locomotive smoke was not a large problem in Chicago, and even if it were, electrification would not be a viable solution.[53]

The association study's citywide approach also de-emphasized one of the most important aspects of locomotive smoke: its concentration around busy tracks, particularly those of the Illinois Central which ran through Grant Park. Although just a small part of the city's total smoke problem, the smoke from locomotives along the lake front had been at the center of the Anti-Smoke League's crusade for electrification. Antismoke activists, as they continued to watch filthy Illinois Central locomotives run through the park, might use their non-expert observation to determine that railroads were still the most offensive polluters in the city.

Perhaps more important, the study did not attempt to estimate the total costs of smoke. In a brief discussion contained in just 17 of the report's 1,052 pages, the study identified several of smoke's negative effects on human health, vegetation, and property within the city, but it did not attempt to determine the costs of these effects to Chicagoans.[54] What the study did do, in great detail, was estimate the cost of electrification. The railroads supplied extremely detailed information to the commission, on everything from switching to overpasses, and the study's experts could easily identify the costs of equipment, new locomotives, transformers, wiring, etcetera.[55] The debit side of electrification offered concrete and reliable figures. But the committee could not so easily quantify the benefits of electrification, both to the railroads and, more important, to the city at large. Still, with only the costs of electrification fully understood and the costs of the continuing smoke all but ignored, the authors of the report comfortably advised against immediate action. After concluding that "pure air is essential to the health and comfort of an urban population," the report advocated only further study of railroad smoke. In this fashion, the report emphasized the costs of environmental improvements over the costs of continued environmental degradation.[56]

After the publication of the association's report, the committee's chief engineer, William Goss, undertook a campaign to publicize the results. In December Goss, who had previously served as the University of Illinois's dean of the college of engineering and as president of the American Society of Mechanical Engineers, spoke before the Franklin Institute's engineering section in Philadelphia, offering an extensive summary of the study's findings. Proud of the study he had guided, Goss was eager to reveal his scientific results to an engineering community deeply engaged with the smoke issue. While lay reformers had reached for a quick fix, a magic cure for the smoke problem, electrification was not the answer, said Goss. The smoke abatement dollar would be better spent in other endeavors.[57]

Goss was not alone in publicizing the findings of the report. Numerous national periodicals paid considerable attention to the report, with some offering detailed summaries and often additional commentary. The editors of *Railway and Locomotive Engineering* proved so supportive of the study that the publicity department of the Pennsylvania Railroad reprinted its commentary as a pamphlet in early 1916. Entitled "Where Smoke Really Comes From," the editorial used the report's smoke figures to conclude erroneously that if electrification eliminated all the railroad smoke in Chicago, "the air would be no clearer, or at least the difference would not be visible to the naked eye." Obviously moving beyond the findings of the Chicago committee, the editor argued that "the idea of electrifying the railway terminals in Chicago is little short of an absurdity." As this effort of the Pennsylvania Railroad's publicity department made clear, the railroads had an important ally in the association, and the committee's report provided crucial ammunition in the electrification debate.[58]

Ironically, just as the Pennsylvania Railroad praised the committee's findings against electrification in Chicago, its own electric technology was winning international attention in San Francisco, at the Panama-Pacific International Exposition. There, at what one author called a "shop window of civilization," mounted in the center of the Palace of Transportation, rotating on a massive, electrically operated turntable, sat the show's Grand Prize, the 4,000 horse-power, 650 volt, direct-current electric locomotive, built by Westinghouse and operated on the Pennsylvania lines into Manhattan's Penn Station. Around the electric locomotive sat the also-rans, an oil-burning locomotive, a twelve-cylinder automobile, and a passenger-carrying airplane, each of which in some form, in relatively short order, would come to dominate various transportation sectors in America, far outdistancing electric locomotion. Even as experts were judging electrification in Chicago infeasible, the judges in San Francisco marveled at the latest electric traction technology. The judgment in Chicago would prove much more meaningful for the future of rail transportation.[59]

Although the Chicago *Tribune* editorialized that the association's report need not guide the city's policy, it did. With the railroads continuing to discuss the possibility of electrification, on their own terms and at their own pace, and with the Illinois Central in negotiations with the city concerning land transfers in and around Grant Park, which would eventually lead to the construction of the Field Museum and Soldier Field, the city further postponed the effort to force electrification. Not until the passage of the Lake Front Ordinance in 1919 did the city council finally legislate on the electrification issue. The 1919 ordinance allowed the city to reach agreement with the Illinois Central to remove all grade crossings downtown and electrify its suburban passenger lines, which it did by 1927, nineteen years after Sergel and the other women of the Anti-Smoke League had initiated their movement, and long after the women had lost most of their influence in the entire matter.[60]

In essence, the Chicago railroads had defeated the efforts of environmental activists to force a specific pollution control solution upon their industry by turning to experts and re-creating the rhetoric along technical and economic lines. With city officials turning to the association's expert report for guidance, the railroads asserted considerable control over the definition of the smoke problem. In the end, the experts concluded that Chicago could not yet afford the improved beauty, health, and cleanliness that members of the Anti-smoke League knew immediate electrification would bring.

In another city, Birmingham, Alabama, which unlike Chicago had no history of activism and no organization dedicated to abatement, businessmen took a similar tack, publicizing economic arguments against abatement in an effort to prevent broad public support of that city's ordinance. They also invited engineers to offer support to their argument that total smokelessness was simply impossible, which was certainly true, making the ordinance appear impractical and harsh. Most important, opponents of the smoke ordinance pitted environmental improvement against economic growth, claiming that the former could be had only at the expense of the latter. In the young, aspiring industrial city of Birmingham, these tactics proved very effective.

As an industrial city, Birmingham shared many characteristics with Chicago. Of course it was much younger and smaller than Chicago, but the rapid growth of the iron and then steel industries after 1880 led boosters to call Birmingham the "Magic City." This southern metropolis owed its growth to the close proximity of iron ore, coal, and limestone, the three ingredients for steel. Indeed, low transportation and labor costs gave Birmingham a competitive advantage over northern mill cities, an advantage defeated by U.S. Steel's infamous Pittsburgh Plus system of pricing, which allowed Pennsylvania steel to compete favorably with the cheaper southern steel. Birmingham was not destined to dominate the nation's steel market, but steel and its major components dominated that city's economy. In 1900, half of the city's wage earners worked in foundries, mills, and machine shops, a percentage that changed little over the next two decades. In addition, thousands of men worked the county's mines, supplying the ore and coal. By 1910, the mills in and around Birmingham made steel the most valuable industry in Alabama, ahead of lumber and cotton. They also made the city remarkably smoky.[61]

In 1907, commercial interests in Birmingham's rapidly developing downtown began to complain about the smoke. In that year, the Birmingham News reported that smoke caused $3 million in damage to merchandise and property annually. The Sloss-Sheffield Company, which operated mills directly adjacent to downtown, took the brunt of the criticism, particularly for its highly polluting coke ovens. By 1911, forces within the city had gained the support of city commissioner James Weatherly, a former railroad attorney, and in late 1912 the city passed its first smoke ordinance.[62]

The new law did not have broad support within the business community, however, and within months many of the city's prominent businessmen launched a sophisticated, calculated attack on the environmental regulation. With pressure from two newly appointed smoke inspectors and the prospect of continuous fines, fifty firms united in a public relations effort against the ordinance rather than an engineering effort against the smoke. Throughout a two-month crusade against the new ordinance, the popular daily newspaper, the Birmingham *Age-Herald,* kept the issue before the people and loyally supported the businessmen's attack. The paper diligently published the complaints of owners who suggested the new law might force them out of business or out of town.[63]

Shortly after the passage of the ordinance, Weatherly offered amendments to reduce opposition to the new regulation, including an extension of the time allotted for starting fires from three minutes per hour to six, during which stacks could smoke without violating the law. This provision would have been very important for industry, since fires required regular fueling and cleaning, both of which caused a temporary cooling in the furnace and thus a spike in smoke production. The six-minute smoke allowance each hour would have made it easier for firemen to conduct these necessary operations without the risk of arrest.[64]

For many of the city's major firms, however, such concessions were hardly enough, and in late January leading manufacturers submitted a petition to the board of commissioners asking for a repeal or major revision of the smoke ordinance. Although officers from distilleries, lumber mills, cotton gins, and a number of other industries signed the petition, most of the signers represented Birmingham's principal industry: steel manufacturing. The signers of the petition argued that the ordinance would actually prevent new businesses from locating in the city, while perpetrating an injustice on those already there. The petition also included one significant detail: "the normal pay rolls[65] of the signers of this petition amount to $2,705,000 annually." Clearly the battle against the smoke ordinance would be waged in economic terms.[66]

Opponents used several arguments against the ordinance. Some claimed smoke was necessary, that burning Alabama's coal required some emissions. Others argued that smoke was good for the city, even suggesting that it was smoke that brought labor and capital to the city and that the elimination of smoke would mean an elimination of economic vitality. Still others emphasized that abatement would be too expensive. One signer of the petition, Donald Comer, secretary of the Avondale Mills company, expressed concern over the cost of abatement equipment. Claiming that the installation of a new

8. The History of Smoke in Alabama. Opponents of abatement often said that smoke brought wealth. This cartoon in a Birmingham newspaper formed part of a campaign to overturn the antismoke law in that city. In the bottom panel, an old fogey makes a fuss before the men assembled (managers and labor alike). "And after the workers had builded a city and factories around the smoke[,] along came the ninny · and demanded that the fire be quenched!" Birmingham *Age-Herald*, 2 February 1913.

automatic stoker would cost his company up to $20,000, Comer labeled the new ordinance a "hardship" on existing manufacturers. Comer also emphasized that no new equipment would eliminate all smoke, due to the poor quality of Alabama coal.[67]

The opponents of the ordinance also very effectively argued that rigorous smoke abatement would mean economic disaster for the city. Through the month-long campaign against the law, the *Age-Herald* attempted to equate smoke with payrolls, using that very term repeatedly when discussing the topic. "Birmingham is a manufacturing city and that means pay rolls and—smoke," noted one editorial. Just two days later a lengthy article announced that the exodus of businesses from the city had begun. D. B. Dimmick, vice president and general manager of the American Casting Company, and also a signer of the petition, argued, "This is an aesthetic proposition—a question Birmingham might well take on some years later, but not while it is a young, struggling city. At the present time we need all the manufacturers, coal mines, and other pay roll producers which we can get: the more the better." Even a Birmingham physician who claimed that smoke was not injurious to health mentioned that he liked "pay rolls better than air without smoke." If the city's residents had not yet gotten the message, *Age-Herald* editors tried to make sure, concluding, "Certainly this is no time for an ordinance unless the people of Birmingham desire to see grass growing in the streets."[68]

The *Age-Herald* was not alone in its opposition to the law. The *Labor-Advocate* also endorsed its repeal, noting that "any hardship that is worked on the manufacturing plants in this district will react on the district itself in running some of the plants off and preventing new ones from coming." Interestingly, the editorial offered a new angle on the smoke debate in proclaiming, "There IS a smoke nuisance in this city which the commissioners can do some thing with, and that is having a crowd of filthy cigarette fiends puffing away in our street cars and other public places for the inconvenience of the passengers." According to the *Labor-Advocate*, the city benefitted from the smoke of industry, but not from the smoke of cigarettes. "Let us try and secure more manufacturing plants, even if we do have to breathe a little more smoke of that character," the editorial concluded, "but let us reduce to a minimum the second-handed inhalation of tobacco and cigarette smoke and also remedy the dust nuisance."[69]

The power of economic arguments rested in the public's general understanding that they had at least some basis in fact. Abatement would be expensive, since it would certainly require extensive investment in new furnace and boiler equipment. Abatement could not be complete, as the ordinance demanded, given the poor quality of Alabama coal and the limited nature of abatement technology. This was particularly true for the beehive coke ovens, which produced dense, damaging smoke while creating the fuel necessary for steel production. A real possibility existed that industries might move to less regulated locations to conduct their fuel-intensive operations.[70]

Birmingham's industrial leaders could argue that the city was engaged in a struggle with its much larger northern competitors, with the Pittsburgh Plus system serving as a constant reminder of the city's inferior status, but at the time of the debate Birmingham was booming. Even as the *Age-Herald* warned of job losses, the Sloss-Sheffield Company announced plans for $500 million in upgrades and extensions of its operations, including opening a new mine in the area, and the Tennessee Coal, Iron, and Railroad Company reported breaking production records at its massive Ensley steel mill. In the face of such positive news, warnings concerning the smoke law's economic impact might have been less powerful.[71]

In case the economic arguments had not had the effect the smoke ordinance foes desired, the *Age-Herald* mustered several experts willing to give testimony against rigorous abatement legislation. One headline read, "Expert Engineers Tell Why Smoke Ordinance Is Not Practical Plan." If readers had not been convinced that the smoke ordinance meant economic disaster through shrinking payrolls, particularly from the mouths of self-interested businessmen who had much to gain from the dismantling of the ordinance, then perhaps the objective testimony of experts might sway public sentiment. The *Age-Herald* quoted Franklin Keiser, a Birmingham-based engineer, who spoke with a voice of authority: "The smoke ordinance is not practical."[72]

The *Age-Herald* also mustered the comments of physicians who denied that smoke affected human health. Since support of the ordinance rested largely on the argument that smoke caused health problems, the testimony of health experts carried considerable weight. The same physician who declared that he liked payrolls better than pure air also stated conclusively, "Smoke is not harmful nor injurious to health." Another physician explained that smoke could not be harmful to health because it did not carry germs. He claimed the intense heat in the fire box purified the carbon, turning on its head the argument that soot was impure simply because it soiled the city. In actuality, this doctor argued, soot was pure, cleansed by the fires of industry.[73]

Even as the city considered amendments that would make the smoke ordinance much more lenient, the *Age-Herald* maintained its attacks. In an attempt to ridicule the smoke ordinance, the paper noted Weatherly's assertion that 91 percent of the city's factories and railway engines were already in compliance with the law. Although Weatherly no doubt intended his remark to reflect how few polluters actually produced the city's dense smoke, the paper asserted that with this statement Weatherly had claimed that the ordinance had greatly improved the city's air in just one month. The *Age-Herald* then reported the thoughts of six residents from various parts of the city. Each resident indicated that the smoke situation had not improved in the previous weeks, suggesting that Weatherly was out of touch with the reality of the problem. The paper also included the thoughts of W. H. Kettig, the manager of a local firm, who criticized not just the law but those who supported it. "The people who want pure air, free from smoke and other ingredients which go to make up a city, should move to the country and take up farming. Smoke always has been and always will be a part and parcel of the city. Smoke abatement agitators should move to the country, for that's the only way they'll ever get away from city smoke."[74]

In this last sentiment, Kettig was correct. Under intense pressure from important businessmen, two of the city's three commissioners (Weatherly being the third) supported a series of amendments to the ordinance which gradually made the legislation more lenient. The original ordinance allowed smoke to issue just three minutes of every hour. By late February, when the commission finally voted on a new ordinance, the city allowed smoke to issue ten minutes of every hour without penalty, plus an extra thirty minutes of smoking each morning for the first firing of furnaces.[75]

With the ordinance so gutted, Weatherly withdrew his support of the legislation and, as the *Age-Herald* reported, "gracefully abandoned the fight for a smokeless Birmingham." All three of the city's smoke inspectors, appointed under the old ordinance, resigned, claiming that the new law left them with nothing to do. Two days after the lenient ordinance replaced the old law, the *Age-Herald* ran a cartoon of the smoke ordinance laid to rest, a wreath from James Weatherly lay at the grave site. To forestall any future threat of municipal regulation of emissions, major manufacturers in Birmingham lobbied the state government to remove municipal authority over the smoke nuisance, which it did in 1915. In a struggle defined as economic progress versus environmental protection, the city of Birmingham had chosen economic expansion.[76]

The 1910s witnessed the transformation of the movement. In Chicago, the old argument of smoke as a serious threat to health, beauty, and morality had been largely subsumed by an economic debate concerning practicality, feasibility, and efficiency. In Birmingham, arguments concerning health, beauty, and morality could not survive in a city where the economic elite were intent only on economic growth. In both cities the potential and actual costs of securing clean air became the focus of the debate. The women and physicians who had led the early smoke abatement movement around the nation had lost much of their authority to engineers and economists. New discussions made clear that society could only pursue a healthful, attractive, and moral environment when it made economic sense. In the 1910s, the shortcomings of efforts in Chicago, Birmingham, and many other cities, confirmed that private economic interests could preempt public environmental concerns. Modern civilization had settled for what many considered to be uncivilized air. While many cities still struggled in earnest against urban smoke pollution, often with real effect, American involvement in World War I would soon prove the frailty of the smoke abatement movement of the 1910s.

seven

War Meant Smoke

In reply the secretary of the interior wrote Mr. Henderson that
the department was well aware of the economical features of
smoke consuming devices, but that war meant smoke and people should stand the smoke in contributing their "patriotic bit."
— Pittsburgh *Post* summary of remarks by Franklin Lane,
Secretary of the Interior, 1917

Conservation—The act of conserving, guarding, or keeping
with care; preservation from loss, decay, injury, or violation; the
keeping of a thing in a safe or entire state.
— *The Century Dictionary and Cyclopedia*, 1911

As war raged in Europe in the summer of 1916, the American antismoke movement reached new heights in popularity and effect. Nowhere
was the heightened influence of the movement clearer than in Pittsburgh,
where that city's Smoke and Dust Abatement League began to organize a
"Smoke Abatement Week." Inspired by England's Smoke Abatement Society and the 1912 smoke abatement exposition at the Royal Agricultural Hall
in London, the Pittsburgh league hoped that a series of events, mostly lectures, would help focus the city on its still dreadful smoke problem and inspire residents to act. The league did not intend the week to stir public sentiment and force government action, however. Rather, members hoped to
influence smoke producers directly, to persuade individuals to take action
against their own emissions. Thus the week's events emphasized the economic aspects of abatement, particularly the increased fuel efficiency which
generally accompanied decreased emissions. The league expected that continued education would convince polluters that economic self-interest alone
made smoke abatement necessary. As if to reveal the harmony of public environmental issues and private efficiency concerns, the league advertized its
hope that these events would help make Pittsburgh a "clean, sunny, and efficient city."[1]

As Smoke Abatement Week began in late October, the dominance of the efficiency theme and efficiency experts became apparent. The chairman of the organizing committee, O. P. Hood, the chief mechanical engineer at the Bureau of Mines' Pittsburgh experiment station, opened proceedings when he introduced the city's chief smoke inspector, J. W. Henderson. Speaking before an audience of manufacturers and railway representatives, Henderson emphasized the progress already made in smoke abatement and the increased economy and efficiency that came through smoke reduction. The next day, Osborn Monnett, formerly the chief smoke inspector of Chicago and at the time an engineer with the American Radiator Company, spoke before the Chamber of Commerce. Monnett discussed the reorganization of the Chicago smoke department along engineering lines and the savings in coal that Chicago had earned through smoke abatement practices. He also showed motion pictures of locomotives in Chicago and Cincinnati using newly designed Altoona jet blowers, which stopped the smoke when activated. The next day Monnett spoke before engineering students at the Carnegie Institute of Technology, outlining his technical knowledge on the subject of smoke abatement. Toward the end of the week, another engineer, H. H. Maxfield, Superintendent of Motive Power for the Western Pennsylvania Division of the Pennsylvania Railroad, discussed locomotive smoke before the Credit Men's Association, identifying several steps that railroads might take to diminish locomotive smoke.[2]

The three most important speakers of the week, Monnett, Henderson, and Maxfield, were all active members of the International Association for the Prevention of Smoke. Indeed, all three had spoken recently before the association. At its annual meeting held in St. Louis, just a month before the Smoke Abatement Week, Monnett had reported on "Stacks and Breechings Formula and Specifications," and Henderson had examined "Front Fired Stokers." Both reports were highly technical. A year earlier, Maxfield had delivered a speech on smokeless operation of locomotives at the annual meeting in Cincinnati. For these three engineers the association had done more than provide a venue for the exchange of smoke prevention expertise; it had also created the personal networks through which information flowed. Monnett, Henderson, and Maxfield were familiar with each other's work and could recognize each other as bona fide experts in the discipline. In a field now dominated by experts, the I.A.P.S. had done much to identify who qualified as such. By 1916, the Smoke and Dust Abatement League honored the expertise of these engineers and ensured that their voices would lead the educational event.[3]

Clearly, men dominated the Smoke Abatement Week's program. Indeed, no women spoke and no man spoke before a women's organization. But participants did not forget the role women had played in the nascence of the smoke abatement movement or the role they could continue to play. Monnett himself made a special plea to Pittsburgh's women's clubs, noting that they could exert a powerful influence upon the enforcement of smoke ordinances, particularly since suffrage appeared imminent. Still, of all the speakers and all the speeches, only one explicitly discussed the issues raised by female activists around the nation, when Dr. William Holman reviewed the health effects of smoke. Holman, a physician associated with the Mellon Institute's smoke investigation, repeated the results of the Mellon research, emphasizing the role of smoke in pneumonia mortality.[4]

The Smoke Abatement Week received considerable attention from the Pittsburgh *Gazette Times,* which presented lengthy articles each day of the event, summarizing the speeches of the previous day and outlining upcoming addresses. On the second day, the *Gazette Times* published an entire page dedicated to smoke abatement, complete with a banner headline, "Make Pittsburgh Smokeless." Accompanying the lengthy article, authored by the Smoke and Dust Abatement League's own John O'Connor, the university economist who had written parts of the Mellon study, the paper ran several related advertisements. The advertisement of one local natural gas firm, the Manufacturers Light and Heat Company, read "Civic Pride Demands Elimination of Smoke Nuisance," and promised that natural gas would vanquish smoke, ashes, soot, dust, work, and worry. Nearby, the Peoples Natural Gas Company's advertisement declared gas "The Safest Smoke Abater," calling it convenient, economical, and reliable, not coincidentally three characteristics Pittsburghers would more likely attribute to coal than natural gas. The Duquesne Electric Light Service also admonished, "Make Pittsburgh Smokeless" in its pitch, in an attempt to convince businesses to allow its central stations to supply them with clean electric power. (When placing the advertisement Duquesne officials were probably unaware that the Smoke and Dust Abatement League would list their company among the city's worst smoke offenders the very next day.) Stoker companies placed the three remaining advertisements accompanying the article, noting the convenience, low price, and reliability of burning soft coal, and emphasizing the possibility of doing so without making smoke.[5]

These advertisements and the extensive media coverage of Smoke Abatement Week reflected the seriousness with which many Pittsburghers regarded smoke reduction in the fall of 1916. The coverage and the adver tisements also revealed how radically the movement had changed over the previous decade. While Monnett noted the need for a well-informed public for success in this endeavor, Smoke Abatement Week made clear that the opinions of engineers were of foremost importance. The smoke problem was technical, and the solution would be technological, involving either the broad use of natural gas or electricity, or the adaptation of some advanced steam technology, such as mechanical stokers. Efficiency, not health or beauty, had become the rallying cry, in Pittsburgh and around the nation.[6]

Thus, when the United States entered the war in the spring of 1917, efficiency dominated the smoke abatement movement as surely as it dominated so many other issues of the day. As the economy geared up for wartime production, efficiency became the watchword on the homefront, and anti-smoke activists hoped to use the great national demand for economic efficiency toward their ends. Smoke abatement officials around the nation argued that efficiency demanded the reduction of smoke. Joseph Lonergan of New York's department of health admonished smoke producers in 1917, "We believe that as smoke means waste, an emission of dense smoke is an unpatriotic act upon the part of the plant owner. The national crisis into which we are plunged demands the saving of all of our resources and especially of fuel." Across the Hudson in Newark, the well-known and respected smoke inspector Daniel Maloney also emphasized the patriotism of smoke abatement in a seventy-eight-page pamphlet, "In the Interest of Smoke Regulation and Coal Conservation, 1918." According to Maloney, the conservation of coal was "a desirable and patriotic duty to benefit mankind."[7]

Clearly, smoke reformers attempted to position abatement as a wartime conservation necessity, and thereby a patriotic duty. In doing so, they retooled the economic arguments against smoke in an effort to make them most effective during the war. Perhaps nowhere was the wartime economic definition of smoke clearer than in a poster competition sponsored by Pittsburgh's Smoke and Dust Abatement League in 1918. The league offered cash prizes to posters that related smoke production to fuel conservation, supplying for contestants an outline of the problem which focused on the relationship between smoke production and inefficient fuel use. The first-prize poster showed a sword labeled "conservation" tamping out the smoke of high stacks and demanding, "Cut out the waste." Other winners used more patriotic themes, with one beseeching businesses, "Don't Shame Old Glory," and another labeling smoke "The Enemy in the Rear." As the league had planned, these posters clearly identified smoke abatement as an essential part of fighting the war on the homefront. League members could not have anticipated, however, how dramatically the war would alter the conservation issue, or how little attention smoke abatement would receive during wartime conservation efforts.[8]

The conservation arguments against smoke took on greater importance when fuel shortages, caused by the rapid economic expansion and the concomitant rise in coal demand, outpaced the ability of an overworked transportation network to carry the coal from mine to market. In the winter of 1916–1917, as coal stocks dwindled, coal conservation became a national priority, and the shortage gave teeth to the efficiency argument for smoke abatement. In Pittsburgh coal conservation became an important theme in the war on smoke, as the demand for steel surged. "It must never be forgotten," wrote Henderson, "that black smoke means loss of heat and that every unit of heat thrown away is so much aid given to the enemy."[9]

Whereas 1917 had seen coal shortages, in the winter of 1917–1918 many cities faced a real coal famine, and the nation struggled through its first energy crisis. Extremely cold weather had not only increased demand for coal but also decreased the supply by exacerbating transportation problems. Frozen waterways blocked coal shipments and severe winter weather, snow and frozen rain, delayed rail traffic and slowed transhipment. In December 1917, riots broke out when residents of New York, fearful of the cold, stormed coal yards to protest exorbitant prices. In one incident nearly two thousand women mobbed a coal dealership in Brooklyn.[10]

In January 1918, as America attempted to furnish its troops and allies with food and war goods, eastern docks overflowed with supplies, and some ships sat in harbors, fully loaded, but waiting for coal. From Washington, where President Wilson had done his part in conserving the precious fuel by shutting off heat in part of the White House, Treasury Secretary William McAdoo, charged with organizing the overburdened railroads, ordered the Pennsylvania Railroad to haul coal through its electrified Hudson tunnels, through Penn Station, and on to Long Island, where supplies were dangerously low. Later in the month, with the coal situation tenuous, New York City contemplated closing schools to save on coal. But the shortages did not affect New York alone, as all major cities faced some hardship, or at least genuine fear of impending hardship. Short coal supplies in the Midwest had even become the limiting factor in the manufacture of steel, a crucial war material. Clearly the nation had reached a crisis, as the great war machine teetered on the brink of paralysis, starving for fuel.[11]

The New York *Times* began to report on the status of individual coal ship-

9. In a show of support for cleaner air during the World War, *Power* published winning entries in the Pittsburgh Smoke and Dust Abatement League's poster contest. *Power* 48(1918): 565.

ments headed for the metropolis, often from mine to market, noting tonnage and expected arrival dates. Several of these coal reports concluded with the city's pneumonia mortality statistics, implying that some New York residents were dying due to the shortage of coal heat. Ironically, pneumonia, the disease most closely linked to the consumption of coal in cities, since smoke could increase pneumonia death rates, was now more closely linked to the shortage of coal. Considering all of its implications, the wartime fuel crisis surely provided ample evidence of the paramount importance of coal in America.[12]

In Washington, federal officials understood the need to take immediate and decisive action in response to the fuel crisis. The shape of the federal response reflected the common understanding of the causes of the crisis. The government took steps in three different directions. First, and by far the most important, it acted to alleviate the transportation logjams which had prevented coal from reaching essential ports and industries. Second, the government took radical steps to create an instant and dramatic drop in the demand for coal. Third, and least important, it attempted to increase the nation's coal consumption efficiency. Notably, smoke abatement played essentially no role in the government plan to ease the crisis, despite the widely held knowledge that controlling smoke could increase fuel efficiency.[13]

As government officials and overworked miners attempted to make clear, the nation could not blame the fuel shortage on the coal industry. Throughout the war the mines produced record amounts of coal, more than the railroads could carry. The New York *Times* concluded, "[I]t is not from a coal shortage which the country suffers, but from a car shortage." With limited storage at mine mouths, some miners even sat idle while awaiting the arrival of rail cars for loading. Federal officials understood that better coordination of the coal-carrying railroads was the most urgent need of the nation.[14] Under the authority of the Lever Act of 1917, the federal government took up this task, creating the Fuel Administration, which in turn established a lengthy set of regulations for coal transportation. The Fuel Administration, headed by Harry Garfield,[15] not only fixed coal prices, preventing coal companies and dealers from gouging during the shortage, but it also established production and consumption zones. Before the creation of the zones, railroads and ships quite commonly cross-hauled coal. For example, coal headed west from Pennsylvania's anthracite district might actually pass bituminous coal headed east from the Pittsburgh district on the same rail line. Some coal traveled hundreds of miles to markets, passing active mines along the way, and occupying valuable coal cars for the extended trip. The Fuel Administration's zones put an end to these long trips, confining anthracite to the east and low-quality Illinois lump coal to the Midwest. The end of lengthy cross-hauls meant not only a savings in total coal car miles, and therefore increased the number of loads each car could carry to market, but also a reduction in the amount of fuel locomotives consumed while delivering coal from the mines.[16]

Although radical in its implications, government regulation of the railroads to alleviate the shortage drew relatively little negative comment from a nation starving for coal. But some of Garfield's other actions did draw significant criticism, particularly those intended to force a dramatic decrease in the demand for coal. The most radical of Garfield's orders involved a five-day industrial shutdown in mid-January 1918, at the height of the coal crisis. Garfield argued that a total cessation of production east of the Mississippi would free rail cars, clear jammed rail lines, and allow waiting steamships to fill their coal bins for the trip to Europe. Garfield also established "heatless" Mondays, during which industry would consume no coal, to occur through March. Although extreme, Garfield's plan did alleviate the rail logjam and went a long way toward ending the coal shortage.[17]

In addition to the fuel zones and the one-time industrial shutdown, the Fuel Administration also established a series of fuel-saving regulations designed to limit coal demand from nonessential segments of the economy. In some cities, operating through its Bureau of Conservation, the Fuel Administration banned unnecessary outdoor lighting, ordered commuter railroads to skip alternating stops, and elevators to stop only above the third floor, all in the name of coal conservation. The Fuel Administration even ordered the curtailment of production in nonwar industries, including tile and brick manufacturing and beer brewing.[18]

The Fuel Administration's least urgent and least effective actions involved the promotion of efficient coal consumption. The Bureau of Conservation sent advisors into factories to hunt for fuel inefficiencies, including unnecessary lighting and idling machines. The government also lobbied proprietors to purchase electricity from central power stations, which could supply power much more efficiently than many older, isolated steam plants. It even arranged for the replacement of carbon filament lamps with the more efficient tungsten lamps, a switch that would be nearly complete by war's end. Still, despite these and other steps toward efficient use of coal, the Bureau of Conservation's actions were slow and had limited effects. Several plans for efficiency were scrapped when the armistice made them unnecessary, including plans to improve the operations of ceramic furnaces and sugar refineries. The difficulty in discerning inefficiencies and devising improvements and the very slow return in coal savings helped ensure that drives for efficiency would play no great part in remedying the coal crisis.[19]

The Fuel Administration also conducted an extensive educational campaign to encourage conservation and to instruct homeowners and businessmen in how to limit their own coal demand. January 20, at the height of the coal crisis, became "Tag-Your-Shovel" day, when school children all around the country brought home tags with instructions on how to burn coal efficiently to put on their family's coal shovels. In addition, the Fuel Administration published thirteen million copies of a pamphlet entitled "Coal Saving in the Home," which also contained hints on efficient home coal use. To educate businessmen, the Fuel Administration worked with the Bureau of Mines to reprint and publicize several of the old bulletins concerning efficient firing of coal plants. The Fuel Administration even encouraged Lester Breckenridge, whose study of smokeless consumption of Illinois bituminous had been so influential, to publish an outline for a course in fuel conservation. Breckenridge responded with eighty pages of lecture notes on the short- and long-term means of coal conservation, including the development of hydroelectric power.[20]

Significantly, Breckenridge, a man who had dedicated so much of his last fifteen years to smoke abatement research, gave the topic almost no attention in his proposed course. In his concluding lecture Breckenridge listed thirteen means of coal conservation, most of which involved reducing waste in mining and preparation, which had long been a major concern of those interested in the longevity of the nation's coal supply.[21] Breckenridge also suggested extensive electrification in cities, recommending the electrification of locomotives specifically. He did not mention controlling smoke in his conclusion. This omission was not unusual. The Fuel Administration did not emphasize smoke abatement in any of its programs or publications. Although conservation and efficiency concerns played an important role during the war, smoke abatement played almost no role in the efficiency campaign, despite the smoke reformers' argument that it must.[22]

One of the government's most important actions toward conservation came not from the Fuel Administration but from Congress. After years of pressure, Congress finally passed the Daylight Savings Bill, modeled after similar actions taken by several nations in Europe. Under the new law, clocks moved up an hour during summer months, so that workers would have extra daylight in the evening. Britain and other nations had already reported great savings in coal and other fuels due to the extra daylight, since demand for artificial light dropped significantly. Although decreasing fuel consumption was the most important argument supporting daylight savings, proponents also offered a variety of other arguments, some credible, some not. Numerous proponents argued that food production would increase, since workers would have more time to garden in the evening when they got home. Others suggested productivity would increase since laborers would be happier with an extra hour of lighted recreation after work. The *Survey* even reported that Britain had seen an improvement in "general moral tone" among its citizens due to the salubrious effects of the extra sunlight and outdoor recreation. Still, the conservation of coal remained the driving factor in the passage of the Daylight Savings Law.[23]

Even as the nation concentrated on coal conservation, however, the meaning of conservation itself changed radically. Prewar conservation methods had involved the promotion of gradual improvements in consumption practices, particularly through the investment in new, more efficient equipment. Conservation was a long-term project designed to improve the economy in the short run and preserve the nation's fuel supply for the long run. But the fears of a fuel shortage in one or two hundred years became quite meaningless in the face of the current crisis; the gradual approach to conservation meant little to cities in immediate need of coal. Instead, the fuel crisis demanded radical conservation. During the war, the issue was not how long the nation's coal fields would last, but how long the coal in the cities' bins would last.[24] Like other prewar methods, smoke abatement was a long-term, gradual approach to coal conservation. Its benefits to coal consumers were relatively small, particularly when compared to the profitability of production during the wartime boom. Firms lost interest in trimming a few percentage points off their coal bills. In the crisis, smoke abatement lost its currency for businesses seeking efficiencies in operation.

Despite the shortages and the emphasis on conservation, coal consumption reached new levels in 1917 and 1918, and the air quality in American cities declined dramatically during the war. In many cities economic expansion increased smoke production, as in Birmingham, where wartime demand forced Sloss-Sheffield to relight its central coke ovens in 1917, just two years after the city had bargained for their closure to reduce smoke emissions. Pittsburghers lamented the return of smoke, as thick as any previous years, as did residents in St. Louis and Cincinnati, cities where smoke abatement movements had made significant progress in the years before the war. Several observers claimed Chicago's smoke cloud had never been thicker. In New York the shortage of anthracite forced many heavy consumers to switch to soft coal, re-creating the smoky situation produced by the anthracite strike in 1902. Even Boston, a relatively clean city, saw a significant deterioration in air quality. But cities with energy-intensive heavy industries, particularly related to metals, suffered the worst fate. Salt Lake City, for example, grew rapidly during the war, and increased activity at the area's enormous copper smelters and expanded railroad traffic led to serious air pollution problems. Salt Lake City's valley location made it prone to temperature inversion, exacerbating its smoky plight.[25]

The increase in smoke was not simply commensurate with the mere increase in coal use, however. Milwaukee's Smoke Inspector Charles Poethke, admitting in late 1918 that his city's smoke nuisance had never been worse, attributed the thick smoke to the Fuel Administration's fuel zones, which prohibited the shipment of semi-smokeless Pocahontas coal to the city. Even homeowners and apartment buildings had been relegated to low-grade Illinois and Ohio coals. The smoke regulations had become meaningless, explained Poethke, who advised the city to "grin and bear" the upcoming smoky winter. "As for health," Poethke reassured city residents, "I expect the smoke in the atmosphere is more disagreeable than unhealthy, for we are only having now what Pittsburgh and Cleveland and some other cities had to endure for years." For Poethke, active in the antismoke movement for nearly two decades, smoke had become a purely economic issue, even to the point where he could dismiss his own previous concerns for the health effects of smoke.[26]

In Philadelphia, Smoke Inspector John Lukens could only watch the smoke clouds gather and offer some explanation to the city: "Many of our large manufacturing plants, on account of the war, have been compelled to force their boilers to the utmost capacity. While this has not been entirely economical and has caused some smoke, the necessity of the case has warranted it and we have, therefore, tolerated these violations, whereas in normal times we would insist upon an abatement taking place." Lukens had hit on the very reason why conservation rhetoric had all but ignored smoke prevention during the war. With industries overworking their equipment, forcing their power plants to operate beyond capacity, and with railroads struggling to meet schedules, pulling heavier than normal loads, proper firing and efficient operation of boilers simply had to await more favorable circumstances.[27]

The rapid deterioration of air quality reflected more than just the increase in economic activity and less-than-perfect boiler room conditions, however. Federal policies designed to alleviate the coal shortage greatly exacerbated the smoke problem. Rather than improve air quality through the promotion of conservation techniques that would reduce smoke, the Fuel Administration inadvertently heightened the smoke crisis through the zoning of coal shipments. As Poethke noted, the Fuel Administration zoned Chicago, Milwaukee, and other midwestern cities out of access to high-quality coals, both Pennsylvania's anthracite and West Virginia's semi-smokeless Pocahontas bituminous. Although most midwestern consumers used soft coal from nearby states before the war, the increased dependence on dirty coal only made the situation worse for Chicago, Cincinnati, Milwaukee, St. Louis, and dozens of other regional industrial cities.

Even in cities that could receive anthracite, such as Philadelphia, the shortage of the prized fuel left many commercial consumers with a choice between switching to soft coal or closing down. Much of the success of Lukens's smoke abatement campaign before the war involved moving soft coal consumers back to anthracite. When the anthracite supply gave out, the city was completely unprepared, for few businesses had invested in equipment that could burn the dirty coal efficiently and with little smoke. Lukens's strategy of the previous twelve years collapsed along with the anthracite supply. Indeed, by 1918, Lukens had all but abandoned enforcement of the antismoke ordinance for the war's duration. He was hardly alone. Years later, Osborn Monnett reflected on the nation's antismoke effort: "The campaign was suspended during the World War, when most of the smoke ordinances throughout the country were held in abeyance with the intent of allowing industries to operate unrestrictedly during the time when maximum production was necessary."[28]

Several other factors also contributed to the rapidly declining air quality. During the energy crisis, consumers used any fuel they could purchase. In Cincinnati, a shortage in natural gas forced some commercial consumers to switch back to coal. Thin supplies of fuel everywhere forced consumers to accept inferior grades of coal, and even coal mixed with dirt and slack, which under normal circumstances would not have been shipped to market. Often the vagaries of the fuel market left consumers with coal grades with which they were unfamiliar and for which their equipment was ill-suited. In other cases the Fuel Administration forced consumers to change fuels in an effort to alleviate shortages. The Baltimore and Ohio Railroad in Pittsburgh, for example, shifted away from coke at the request of the district's fuel administrator, who deemed coke too scarce and precious to burn in locomotives. The Baltimore and Ohio had been using coke on its Junction Branch specifically to reduce smoke. The result of such fuel changes was uncommonly dense smoke.[29]

A drain of skilled labor also impeded ongoing efforts of firms to increase fuel efficiency and decrease smoke emissions. As thousands of men enlisted or were drafted for war and thousands of others changed jobs to take advantage of the labor scarcity, many businesses lost key personnel in their boiler rooms and locomotives. Employers found it particularly difficult to find and keep skilled firemen, who could more easily find less strenuous and higher paying work due to the high demand for labor. With less experienced men fueling their fires, firms could anticipate lower fuel efficiency and greater smoke production.[30]

To some degree Secretary of the Interior Franklin Lane spoke the truth when he explained to J. W. Henderson that war meant smoke. The nation could hardly expect to overcome the structural problems that necessitated the restriction of coal shipments or the situational problems that required massive production and the operation of boiler plants beyond capacity. In these ways, the war really did mean smoke.[31] Still, the collapse of the smoke abatement movement and the cessation of municipal enforcement of anti-smoke ordinances signaled a failure much more complete than that required by wartime predicaments. Before the war air quality had become an economic consideration that could be sacrificed, just like meat, wheat, electric street lights, elevators, or any other commodity or convenience given up for the sake of the war effort. Pure air was no longer a natural right but a noble goal worthy of consideration. Unfortunately, during the war, politics necessitated a failure in the protection of clean air.

As if to illustrate the increasing role of private economic issues in the protection of clean air, the International Association for the Prevention of Smoke underwent a transformation during the war. Using the newer name Smoke Prevention Association, the organization first elected an officer from the private sector in 1916, when W. L. Robinson, a fuel supervisor from the Baltimore and Ohio Railroad became Second Vice President. Following his election, a railroad employee consistently filled this post. During the next convention in Columbus, Robinson showed two short films made by railroad companies, one illustrating the effect of brick arches on locomotives and the other on the work of the Pennsylvania Railroad toward smoke abatement at its Pittsburgh yards.[32]

In 1918 the association took a dramatically commercial turn at its thirteenth annual convention in Newark that fall. Since changing its rules of membership in 1913, allowing persons who were not government smoke officials to join, the association had not become a popular organization for lay reformers, as some had hoped. Instead, business representatives had joined, to the point that in 1918, despite the presence of Newark's Maloney, businessmen dominated the organization. The professional smoke officials who had previously controlled the organization were only sparsely represented.[33]

Those who attended the Newark convention heard ten presentations, nine delivered by men in the private sector. Engineers from the Underfeed Stoker Company, the Green Engineering Company, and the Locomotive Stoker Company gave lectures, as did officers from the Locomotive Pulverized Fuel Company, the Kewanee Boiler Company, and the International Coal Products Corporation. In each presentation, corporate officers and engineers discussed how their companies' products could help alleviate the smoke problem. As if to further punctuate the corporate dominance of the association, attendants spent an evening watching industrial films, including one provided by the Westinghouse Electric Company concerning the electrification of railroads, and another from the Kewanee Boiler Company, entitled "The Troubles of a Building Owner." Obviously, the Smoke Prevention Association's annual meeting had become an important venue for firms developing abatement equipment or smokeless fuels to publicize their ideas and products. And, during the war at least, it had also become a much less important gathering for government smoke officials hoping to professionalize their field.[34]

If the shifting focus of smoke abatement from public environmental concerns toward private efficiency issues seems natural, ordinary, or even necessary, compare the issues of pure air and pure drinking water. In their efforts to force action toward smoke abatement, early antismoke activists had attempted to relate air pollution to health, often comparing airborne pollutants to waterborne pollutants, which urbanites had already intimately linked to disease. Activists also contrasted the considerable municipal efforts toward supplying pure water, undertaken for health reasons, with the weak efforts to secure pure air. As Chicago Health Commissioner W. A. Evans remarked in 1908, "Pure air is just as essential to a high level of public health as pure water." He and others wondered why cities had spent so much money on securing the latter and very little to secure the former.[35] It seems reasonable to suggest that urbanites would have balked at giving up potable water in an effort to fight a war in Europe, or that municipalities would not have allowed sewage to collect in puddles in city streets while shifting resources toward more immediate war needs. The issues of pure water and sewage removal remained primarily health concerns, and, in fact, two of the most important health concerns facing municipalities. Pure air, on the other hand, had not become primarily a health issue, and in the face of the national war effort pure air could be sacrificed.[36]

The gathering smoke clouds of World War I indicated the failure of a philosophy. The antismoke environmentalism stressing the importance of health, beauty, and cleanliness had given way to an antismoke conservationism stressing economy and efficiency. During the war this conservationism had failed to prevent air quality from deteriorating dramatically. As an economic factor, smoke production had entered the very complex economic equation that governed business decisions; particularly during the war, this equation amplified other factors and greatly diminished the relative importance of controlling smoke.

"Where's My Smoke?"
Movement toward Success

A whole cityful of people cannot be prosecuted. The very idea is
absurd. . . . The solution to the smoke problem will only be
found when it becomes possible for the great masses of people
to buy a smokeless fuel at approximately the same prices they
are now paying for smoky Illinois coal.
>   —St. Louis *Post-Dispatch* editorial, 24 November 1937

We simply cannot tell the small home owner to stop emitting
smoke any more than we can tell an automobile operator to stop
exhausting poisonous fumes into the atmosphere.
>   —Charles J. Colley, Monsanto Chemical Company,
>   16 March 1939

The world war meant only a hiatus for the smoke abatement movement, not its demise. Although the remarkable increase in smoke failed to
rekindle a powerful public movement for environmental reform after the
war, some cities saw their old efforts renewed along familiar lines. In Milwaukee, for example, local newspapers covered several attempts to curb
smoke emissions in the decades between the world wars. The Milwaukee
Journal reported, "Doctors Attack Smoke as Menace to Health" in 1928, and
"Smoke Perils Health of City, Says [Health Commissioner] Koehler," in 1929.
Three years later the paper announced, "War on Smoke Is Begun," and
promised, "Science to Rid City Air of Smoke and Gases." These latest efforts
to control coal smoke rested on old ideas: that smoke was harmful to health
and that new technology would offer the solution.[1] Milwaukee also continued its reliance on engineers to find that solution, accepting the role of experts and the laws of economy in guiding the city toward cleaner air. By 1935,
the *Journal* announced that Milwaukee had surpassed Chicago in smokiness
and that the problem was growing worse. By 1941 the Bureau of Mines esti-

mated Milwaukee County's total loss to coal smoke at $10.8 million annual-
ly. Obviously the renewed efforts along old lines had not produced the desired results.[2]

Cleveland too saw a revival of its smoke abatement crusade, when the
Women's City Club initiated a new campaign in 1922, also attacking smoke
as a health problem. The club published a "smoke coupon" in local papers
which allowed residents to simply fill in blanks to report smoke violations
near their homes. The women received several hundred complaints using
these coupons, which they then passed on to the city. Reviving a more traditional tactic, club members also took to the roof of the YWCA to observe
smokestacks and issue their own complaints. Throughout this campaign the
Cleveland women focused on the old issues of health, beauty, and cleanliness, and after ten months of work, Sarah Tunnicliss of Chicago's Women's
City Club visited to deliver an address entitled, "What Right Has a City
Dweller to Breathe Clean Air?" Clearly the environmental concerns articulated in the 1890s had not disappeared.[3]

Cleveland officials responded to the new pressure brought by the women,
quickly firing the incompetent smoke inspector and eventually establishing
a better-financed Division of Smoke Inspection. By 1926, however, the energy of the Women's City Club had waned, the city's efforts had proved inadequate, and the smoke problem persisted. Eventually Cleveland joined the
long list of cities that again saw their smoke abatement efforts curtailed by a
national emergency, this time the Great Depression.[4]

St. Louis also witnessed an intensive educational campaign against
smoke. Begun in 1923 by the Chamber of Commerce and the Women's Organization for Smoke Abatement, the campaign heightened the public debate and culminated in a 1924 ordinance which expanded the smoke abatement department. But this old-style campaign led by nonprofessionals
brought limited results, and St. Louis, like more and more cities, eventually
placed its faith in a different type of effort. George Moore, director of the Missouri Botanical Garden, summarized the newer approach while speaking before the Women's Chamber of Commerce. "Keen reformers with boundless
enthusiasm and a complete lack of technical knowledge," Moore announced,
"are hardly the ones to convince the manufacturer, the manager of the apartment house or the ordinary householder that smoke is such a frightful menace after all." Moore urged the wholesale replacement of antismoke propaganda from non-experts with technical research and education. In this type
of antismoke crusade the women to whom he spoke and even Moore himself would have little voice. The St. Louis *Post-Dispatch* agreed with Moore's

assessment. In an editorial the paper encouraged the initiation of a new form of antismoke activism. There would be no call to arms, no appeal to mass support for clean air. Rather the paper called for a comprehensive study of the St. Louis smoke problem along the lines of Pittsburgh's Mellon investigation, conducted, obviously, by experts in the field. After thirty years of antismoke activism, research, and education, the *Post-Dispatch* apparently decided it was time to start anew.[5]

As anticipated in the years before the war, the smoke abatement movement continued to evolve from an interest-group-dominated political effort into an expert-controlled scientific endeavor. Perhaps nowhere was the transformation so clear and so complete as in Salt Lake City, where rapid industrial growth during the war had led to a dramatic decrease in air quality. In early 1919, the city commission voted to initiate a scientific study of the Salt Lake smoke problem, and aided by Utah Senator Reed Smoot's efforts to appropriate federal funds, the city engaged the Bureau of Mines to lead the study. The bureau in turn retained Osborn Monnett, then practicing as a consulting engineer in Chicago, to direct the investigation, while Bureau Chief O. P. Hood provided guidance from Pittsburgh. With a sizable professional staff, consisting of federal employees, University of Utah scientists, and local engineering experts, the investigation studied weather conditions, monitored smoke emissions using the Ringelmann Chart, conducted a soot fall study in the city, and published a series of recommendations for Salt Lake City.[6]

Following Monnett's recommendations, Salt Lake City passed a new ordinance in 1920, which created a professional smoke inspection department. The city then hired H. W. Clark, an engineer who had participated in the investigation, to serve as the its first smoke inspector. Clark conducted his department using Monnett's guidelines, emphasizing education and inspection and minimizing enforcement and prosecution. Clark focused his work on upgrading boiler room equipment and accepted the gradual approach to abatement supported by most engineers. As O. P. Hood concluded, smoke abatement was a "long-time effort" requiring a "willingness to wait for results." Salt Lake residents proved more than willing to wait, and in 1926 the city stopped funding the smoke inspection department. Although the rapid increase in air pollution that followed forced the city to partially re-fund the department, the valley's air quality continued to deteriorate until atrocious conditions in the 1940s sparked a new wave of activity.[7]

In Salt Lake City and around the nation engineers dominated the post–World War I discussion of smoke abatement. While lay reformers con-

tinued to press environmentalist arguments against the smoke, the conservationist issues of efficiency and economy received more attention from municipal officials. However, as smoke abatement officials continued to preach
the gospel of efficiency to owners and operators in the 1920s, they yielded
fewer converts. Old arguments concerning the conservation of coal had lost
considerable currency, even after the acute shortages of 1917 and 1918. After
the war, coal prices dropped, and notwithstanding periodic strikes in the
1920s and concomitant spikes in prices, bituminous coal supplies would never again shrink against a swelling demand as they had during the war. The
predicted exponential growth in coal consumption failed to materialize, and
coal prices flattened. With fuel prices low, coal consumers sought economies
in other areas of production, particularly labor costs.[8]

Even as the economic incentive for individual fuel conservation declined
in the 1920s, the national concern for coal conservation also dwindled. In
1908, the Bituminous Coal Trade Association had predicted annual domestic production would reach one billion tons per year by 1935, but it failed to
reach half that amount. Indeed, in that depression year, the nation produced
only 2 percent more coal than it had in 1908. With little increase in demand
for coal and a greater awareness of the potential of other fuels, particularly
petroleum and natural gas, a concern for the eventual exhaustion of coal
seemed premature if not preposterous. In 1917, two government scientists
estimated that the nation had consumed only 1 percent of its bituminous coal
supply. Surely the time for conservation had not yet arrived.[9]

As the salience of coal conservation waned in the 1920s and 1930s, the postwar antismoke efforts proved no more successful than those before the war,
and perhaps even less so. Engineers now dominated the national dialogue
concerning smoke, but, though the dominant technical discussion largely excluded the voices of lay reformers, the persistent clouds continued to incite
public scorn. In addition, the prosperity of the 1920s encouraged activists to
make significant demands, as many reformers assumed that corporations had
investment capital. The railroad electrification debate, for example, heated up
as electrified lines in New York, Philadelphia, and Chicago proved their value, both to city residents and railroad corporations. Without relying on old arguments concerning the efficiency of electric traction versus steam, residents
in several cities heightened their clamor for more extensive improvements
based largely on the benefits accrued directly from smoke elimination.

Not surprisingly, Chicago witnessed an intense debate over electrification
in the mid-twenties, particularly in 1926 when the Illinois Central opened its
electrified suburban line along the south shore. Even before those opening

celebrations, however, the city's newspapers had renewed their campaign for electric traction. The *Tribune* published articles about New York under the headline, "Sootless Air, Beauty Found in Electrifying," and on Philadelphia under, "They Electrify for Beauty Spot in City's Heart." Both articles emphasized not just the aesthetic value of electrification but also the economic. The author extolled the value of "air rights," which became a catchphrase in the debate. Electrification of terminals, particularly Grand Central and Philadelphia's Broad Street Station, allowed railroad corporations to develop land above electrified lines. In New York the broad Park Avenue was rapidly becoming an exclusive address, particularly along the blocks just north of the station, where previously steam locomotives created what the *Tribune* called a "black, belching, fiery furnace."[10]

By the mid-1920s the collection of skyscrapers that surrounded the station provided clear evidence of the success of Grand Central's electrification. Removal of steam locomotives allowed the New York Central to construct streets and buildings on the nearly twenty-nine acres of land above the terminal itself, and increased the desirability of surrounding property. In the fifteen years following the station's opening, the blocks above the tracks between Madison and Lexington avenues from 42nd to 45th streets and several additional blocks north along the restored Park Avenue rapidly became some of the city's most valuable real estate. On this land the New York Central itself constructed several prominent structures, including the Biltmore and Commodore hotels, both named in honor of the road's corporate architect Cornelius Vanderbilt. Following these two magnificent hotels came the Ambassador, Marguery, Park Lane, and Roosevelt in the 1920s, and finally the Waldorf-Astoria, completed in 1932. In addition to the high-rent hotels came high-rent office space, most notably in the Graybar and the New York Central buildings, with the latter literally straddling Park Avenue and facing north over a new civic center. Along the stretch of Park Avenue between 42nd and 96th streets, property values increased 374 percent between 1904 and 1930, and the railroads made millions in rentals for their air rights.[11]

Although perhaps less dramatic than the changes in midtown Manhattan, electrification promised great improvement in Philadelphia's center city. The Pennsylvania Railroad announced plans to electrify its Broad Street Station and bury the "Chinese wall" of steam lines that ran from City Hall to the Schuylkill River. For the task the city was to pay the cost of reconstructing streets, sewers, and gas mains. Philadelphia also continued to reap the benefits of the electrification of the Pennsylvania's line north toward New York and south toward Washington.[12]

Success in the East, particularly New York, assured that electrification would remain a major issue in Chicago until the post–World War II replacement of steam with diesel-electric. Not only did the eastern projects give good evidence of the practicality and desirability of the new technology, but the realization that such developments continued to bypass the great railroad hub hurt the pride of many Chicagoans. The *Tribune,* for example, ran a stinging editorial after the Pennsylvania Railroad ordered ninety-three more coaches for its newly electrified line between Philadelphia and Wilmington. "If electrification is profitable," the *Tribune* protested, "we do not see why we in Chicago should not get some of the gravy." The editorial went on to suggest that the railroads favored the eastern metropoli since they were home to the corporations' directors and presidents. "The rails lead to Chicago but the directors stay in the east. The smoke of their locomotives does not soil their collars."[13]

If the improvements in New York and Philadelphia put pressure on the railroads in Chicago to electrify, the completion of the Illinois Central electric line did more so. Mayor William Dever put it rather plainly while testifying before the city council's committee on railway terminals: "If the Illinois Central can do it, every other railroad can do it." On the day the I. C. began its electric operation, the *Evening American* ran an editorial announcing the "dawn of a new and brighter day for Chicago." "Electrification of the I. C. should be but the start, not the end. Every terminal in Chicago should be electrified." On the same day the *Daily News* ran a political cartoon showing a new I. C. electric motor car walking swiftly past two elderly steam locomotives. "Take a tip from me," the I. C. locomotive says, "I feel one hundred per cent better and have a lot more pep since I've quit smoking." This was not the only cartoon to link locomotive smoke with tobacco. Five days later, the *Tribune* ran a cartoon titled "The Reformed Smoker," in which a young, handsome man representing the I. C. suburban service stood smiling as a gallery of ladies looked on admiringly. In the foreground overweight, older men sat sour-faced and smoking large cigars, with "Unreformed Railroad Smokers" written across their backs. "I feel fine since I cut out smoking," announced the I. C.[14]

The railroads met this agitation for electrification much as they had previous efforts. Railroad officers pleaded before the city council that electrification in Chicago remained too expensive. T. B. Hamilton, vice president of the Pennsylvania Railroad, reminded the council that in no case had "the use of steam service been eliminated due to the so-called smoke nuisance alone," and that the extenuating circumstances in New York and Philadelphia did not exist in Chicago. Railroad officials also suggested that the 1915 Association of Commerce report be brought up to date, a project which might delay action for another year. But as the city council lost interest in forced electrification in the following months, the railroads saw no need to fund the report's updating.[15]

If the agitation for electrification failed to produce the desired results in Chicago, so too did antismoke activism in many other cities, including Cleveland and St. Louis. Most likely any incremental improvements in air quality in the 1920s were dwarfed by the impact of the Great Depression of the following decade. During the 1930s, bituminous coal consumption averaged just 370 million tons per year, down from an average of 488 million tons per year in the 1920s, meaning that the Depression brought a 24 percent decline in soft coal consumption. This decline, particularly within industry, brought some localized improvements in air quality, but obviously at a great price.[16]

Unlike the crisis of World War I, the Great Depression did not bring abandonment of smoke control efforts. Although tight city budgets forced the reduction of smoke inspection staffs and the poor economy greatly reduced support for strict enforcement of laws, in one crucial respect the Depression brought significant advancement. Like so many civic issues, smoke abatement received the attention of federal programs designed to end the economic crisis. Both the Civil Works Administration and the Works Progress Administration funded antismoke projects, some of which became bases for further action.

In early 1934, the C.W.A. financed a smoke investigation in Chicago. Federal money allowed the city to hire 168 unemployed engineers to operate under Deputy Smoke Inspector Frank Chambers. The department used the engineers to cover Chicago like never before, as Chambers stationed at least one new inspector in every city ward. Previous coverage had rarely extended beyond the central business district and the major industrial zones. The project, organized in part by Osborn Monnett, included studies of meteorological conditions, soot fall, fuel consumption, and plant equipment. It also allowed the engineers to report violations of the law, which they did 4,559 times in the three months of the project. The engineers gave instructions to furnace operators in 7,279 instances, and oversaw the reconstruction of 269 plants.[17]

Chambers led a publicity campaign against smoke as he introduced the new C.W.A. project. He used traditional outlets for reform activism, newspapers, and association meetings, but he also used the radio to introduce the program and to increase smoke abatement awareness generally. For example, on January 23, the Chicago Federation of Labor's radio station, WCFL, broadcast a lengthy report on the project. It began: "The message we bring to you today should be of vital interest to house-wives and mothers as it seriously concerns not only the health of adults and children but has to do with physical cleanliness of the home and the destruction of property." With this broadcast Chambers hoped to appeal to the traditional supporters of anti-smoke activism, women, and he did so using very traditional complaints against the smoke. The broadcast went on to describe the C.W.A. project and directly linked smoke with several health problems, including "respiratory diseases" and a lowered "general tone of the community."[18]

As might be expected, certain segments of the city were none too pleased with 168 new smoke inspectors on duty. One Pennsylvania Railroad official noted that since smoke would inevitably issue from steam locomotives, the increased attention "looks very much like another electrification pressure scheme." But the smoke project in Chicago did not outlast the short-lived C.W.A., and its replacement under the Works Progress Administration operated under different guidelines. The W.P.A. project, an "Air Pollution Survey," employed over five hundred men engaged in research and education, but not enforcement. Indeed, to ensure the cooperation of plant operators and railroad officials Chambers explicitly stated that the W.P.A. engineers had no law-enforcing ability. Chambers apparently had long-term goals in mind, specifically gaining a fuller understanding of the smoke problem, its sources, and causes.[19]

Chambers shared this goal with several municipal officials who used W.P.A. funding to conduct smoke surveys around the nation, including Health Department officials in New York. The New York survey included a study of fuel consumption, soot fall, and distribution of fuel-burning equipment. A preliminary report of the survey appeared in the *American Journal of Public Health* in the spring of 1937, announcing that the data collected by the survey would "form the ideal background of information upon which an intelligent campaign of smoke elimination can be based." Unlike most of the W.P.A. smoke projects, the New York survey ended in a published final report, with the data appearing in *Heating, Piping, and Air Conditioning*, the journal of the American Society of Heating and Ventilating Engineers.[20]

Other surveys published their data in free-standing reports, including the massive, unpaginated, "Air Pollution, City of Pittsburgh, PA." This report included the data gathered by W.P.A. engineers from December 1937 through March 1940, including soot fall reports from one hundred collection stations around the city. Still, like most W.P.A. surveys, the report offered neither a summary of its work nor conclusions drawn from the data. The Pittsburgh study did differ from others, however, in the inclusion of a health questionnaire, entitled "Smog and You," to which the Department of Public Health received nearly 4,500 responses.[21]

The failure of the Pittsburgh survey to offer conclusions or policy recommendations is not surprising given the nature of the program. As the Cleveland Smoke Abatement Project announced, the purpose of the W.P.A. surveys were twofold: "first, to provide employment to technical and clerical persons; second, to conduct an accurate survey of soot and smoke conditions in the City." Once the war made the first aim unnecessary, many projects were left incomplete, often with their sole public reports issued mid-stream.[22]

Their unfinished nature should not suggest that the projects had little impact, however. In Cleveland, for example, the project involved training over thirty engineers in smoke abatement techniques, including proper use of the Ringlemann Chart and soot collection jars. Another 225 men received training in plant inspection. Thus, in addition to the raw data, which in Cleveland included the location of all industrial heating and power plants, the projects provided education to engineers and the public alike. W.P.A. funding also allowed cities that had exerted very little effort toward smoke abatement to conduct extensive surveys. This was the case for Louisville, where in late 1938 twenty-eight engineers conducted a "Smoke Abatement Study of Industrial Plants and Railroads," giving that city a clear picture of its air pollution problem for the first time. Even the Pittsburgh survey provided important data for antismoke activists. Just a year after its completion, the W.P.A. study drew the attention of the Civic Club's smoke elimination committee. Here, an important figure in the movement, Jay Ream, stressed the value of the report's implication of residences in the problem. Ream noted that 120,000 Pittsburgh residences burned soft coal, arguing that any solution must address this smoke source. The W.P.A. studies in other cities, including St. Louis and Louisville, also pointed city officials toward domestic sources as major culprits in their smoke problems, providing significant impetus for changing the approach toward a solution. Indeed, with industry crippled by the Depression, the most important legacy of the W.P.A. studies may have been their focus on domestic sources. In several cities these reports predicted, and perhaps encouraged, a growing concern regarding smaller sources of pollution.[23]

The importance of the W.P.A. surveys in keeping the issue before the public during the Depression and in developing more reliable data on the extent and sources of the problem underscores the centrality of expertise in the search for a solution. However, the ad hoc groups of W.P.A. engineers cer-

10. Darkness at Noon. Smoke makes visibility poor on this Pittsburgh street in the
1920s or 1930s. Courtesy of the Carnegie Library, Pittsburgh.

tainly did not represent the avant-garde of the smoke control profession.
Through the 1920s and 1930s the Smoke Prevention Association continued
to represent engineers and other professionals dedicated to the issue, and it
remained the most important locus for building professional identity within the field of smoke control. The membership included a mix of municipal
smoke inspectors and corporate officials. Although inspectors always held
the president's post, and Frank Chambers of the Chicago Smoke Inspection
Department held the secretary-treasurer position through both decades, by
the mid-1930s private sector influence within the leadership of the association had grown. Railroad officials consistently filled the second vice president's post, as they had since its creation in 1916, and after 1928 railroad men
also filled the new position of sergeant-at-arms. Perhaps more important, in
1934 coal industry officials first appeared among the ranks of association officers, when W. E. E. Koepler of the Pocahontas Operators Association became first vice president, a post he maintained until Stanley Higgins of the

New River Coal Operators Association replaced him in 1938. Koepler's office helped ensure that low-volatile coal interests would have a voice within the association, but high-volatile interests had no comparable voice.[24]

Continuing a trend begun during World War I, men in the private sector also gave a preponderance of the talks before the association. At the 1939 convention industry representatives gave seven of the fifteen papers, while municipal officials gave just two. (The remainder came from three academic scientists, two federal government officials, and a lawyer for the Indianapolis Smoke Abatement League.) The next year eleven of the eighteen speakers came from the private sector, with seven representing the coal and railroad industries. Engineers moved easily between municipal, academic, and industrial employment, creating what historian David Noble has called a "professional-corporate-educational community of engineers." Among the engineers spending significant time in both the private and public sectors were Paul Bird, who moved from his post in Chicago's antismoke department to Commonwealth Edison, and Osborn Monnett, who worked in several government positions and for several businesses during his long career. A common professional interest in abating smoke through technology, whether approached through government or business employment, outweighed any seeming conflict represented by the regulator/regulated divide. In other words, engineers' loyalties to their profession lay across any boundaries suggested by public service in regulation of industry. The Smoke Prevention Association provided just one of many venues where engineers with similar engineering interests but very different occupational goals exchanged ideas.[25]

Although the Smoke Prevention Association annual meetings remained an important site for dialogue, the most important innovations within the movement came not from the association but from a campaign waged in St. Louis in the late 1930s. Beginning with the election of Mayor Bernard Dickmann, and his subsequent appointment of Raymond Tucker as his personal secretary, St. Louis took the lead in smoke abatement efforts. In 1934 Dickmann instructed Tucker, a professor of mechanical engineering at Washington University, to "clarify the air as a previous administration had clarified the water." Tucker accepted the challenge, understanding that the problem required a new approach. As he well knew, St. Louis had witnessed periodic organized smoke abatement efforts since 1893, with the latest push coming after the 1923 founding of the Citizens' Smoke Abatement League by the Chamber of Commerce and the Women's Organization for Smoke Abatement. Tucker was highly critical of the educational campaign organized by the league, calling it futile, and he plotted a radical departure from the two failed strategies then commonly employed: strict enforcement at the stack and education at the boiler / furnace. Tucker's strategy represented a third approach: regulation at the coal yard.[26]

Working from the analogy of water purification, Tucker reasoned that the city could not expect all of its citizens to use poor quality coal without producing smoke. Instead, if the city forced the improvement of fuel before distribution, just as it did with water, then individuals could stop concerning themselves with smokiness. Of course, Tucker was not the first to suggest that smoke elimination would await the widespread use of smokeless fuels. Convincing offending proprietors to switch fuels had long been a viable strategy, particularly in cities where alternatives to bituminous were reasonably economical. But in a city like St. Louis, where perhaps 94 percent of its bituminous coal came from the counties just across the river in Illinois, smokeless fuels were simply not economically competitive. Forcing a change in fuel consumption would clearly require the exclusion of the cheapest (and dirtiest) coals or the processing of such coals into more a acceptable form.[27]

After a Dickmann-appointed citizen's committee advocated the restriction of soft coals, Tucker hired the omnipresent Osborn Monnett to study the St. Louis situation and to offer recommendations. In December of 1936, Monnett recommended that large fuel consumers, industry and apartment buildings, use only mechanical stokers, and that smaller consumers, including homes, switch to smokeless fuels. Even while suggesting that the largest part of the problem lay in domestic consumption of soft coal, Monnett, like almost everyone speaking on the issue, conceded that any solution must include the continued use of southern Illinois coal. Tucker followed Monnett's recommendations in writing a new smoke bill, but he also made one important addition: a "washing clause," which required the removal of impurities from coal sold in the city. This provision could reduce sulphur and fly ash in fuel, but as opponents argued, it could hardly reduce the smoke problem.[28]

As the city council debated the bill, Illinois business interests campaigned against any provision that would harm the coal industry. In early 1937, the Belleville and Collinsville Chambers of Commerce wrote an open letter to St. Louis industry which appeared in the *Post-Dispatch*. After claiming that the proposed law would double the cost of the city's inexpensive fuels, the letter invited St. Louis industries to relocate across the river in St. Clair and Madison counties. Illinois coal men also argued that the washing clause would force the closure of mines lacking the capital to construct and operate a washing facility. Even the *Post-Dispatch* opposed the clause, calling it unnecessarily antagonizing of the city's Illinois neighbors. However, unlike the Illinois opponents, the newspaper favored a stronger ordinance, one that could stop the smoke, not just reduce fly ash and sulphur.[29]

The passage of the bill in February brought an unsuccessful coal company lawsuit, which argued that the ordinance represented an unreasonable interference with interstate commerce. The law also brought companion legislation, which followed in October and provided enforcement provisions. This new law regulated the importation, storage, and distribution of coal within the city, and empowered the city to license fuel dealers. By the time the second ordinance had passed, Tucker had agreed to head the smoke division, supported by a mayor dedicated to smoke control and the *Post-Dispatch*, which offered continuous coverage of the issue. Not surprisingly, though, smoke continued to plague St. Louis, particularly in the winter months when dense clouds could slow rush-hour traffic and force the use of streetlights past midday. After two years under the law, Tucker refused to apologize for the continuing smoke, noting instead that both fly ash and sulphur emissions had been reduced. In the meantime, Tucker also used the observations of W.P.A. workers to conclude that homes and apartments contributed a significant share of the city's smoke, suggesting the importance of finding a strategy that could keep the smokiest fuels out of domestic fires.[30]

On Sunday, November 26, the *Post-Dispatch* ran a lengthy editorial entitled "An Approach to the Smoke Problem." More than a call for action, the editorial made definite proposals. After noting that the solution should "not involve the wreaking of the great soft coal industry in Southern Illinois," the editorial suggested that the city enter the fuel market, purchase smokeless coals, and distribute them at reasonable prices. Perhaps more important than the specific recommendations, the *Post-Dispatch* conducted a vigorous anti-smoke campaign over the next two weeks. The issue appeared on the front page for fifteen of the next sixteen days, often accompanied by companion articles and editorials on other pages. Mostly publishing the testimony of prominent St. Louisians and photographs of the blackout smokes that haunted the city in late November and early December, the paper was never short of column inches and attention-grabbing headlines. Among them was a front-page article announcing "Shape of Soot Particles Promotes Cancer of Lung, Tests Indicate." Several of the paper's photographs of November 28, known as "Black Tuesday," later appeared in *Business Week* and *Life* magazines.[31]

In response to the *Post-Dispatch* campaign, Mayor Dickmann invited fifty-two "civic leaders" to convene on the issue. As an indication of how much the new approach had already changed the definition of the smoke problem, the *Post-Dispatch* referred to the meeting as a "Clean Fuel Conference," not a smoke abatement conference. Among those attending were Morton May, president of the May Company, Thomas Dysart, president of the St. Louis Chamber of Commerce, and Dr. Philip Shaffer, dean of the Washington University Medical School. Out of the meeting came a new citizen's smoke elimination committee composed of seven prominent men (and no women), including a real estate dealer, a retired broker, a Monsanto Chemical vice president with a technical background, the retiring president of the St. Louis Medical Society, the mayor's secretary, and Tucker. James Ford, a vice president with First National Bank, chaired the committee. Aside from the Monsanto executive and Tucker, none of the members had engineering experience, and the committee appointed two official research advisors, the Missouri state geologist and the chief of the Illinois Geological Survey. The selection of fuel experts rather than steam engineers gave witness to the changing ideas about where to attack the smoke problem.[32]

Two months after its appointment, the committee issued a report. The men recommended that the city take drastic measures, all of which had been discussed for several years: that all consumers of high-volatile coal use mechanical stokers; that anyone not employing a stoker use smokeless fuel; and that the council empower the city to purchase and distribute coal in emergencies. Although businesses in Illinois coal country briefly threatened to boycott St. Louis products, opposition to the new law within the city was apparently slight. In April, after the city council easily passed the bill, Mayor Dickmann crowed, not without some justification, "This is the greatest single thing we have ever done in St. Louis."[33]

To make the new law work, the city had to secure low-volatile fuels for domestic use at a reasonable cost. For the first winter, the city largely accomplished this through a special arrangement with the Frisco Railroad for the shipment of Arkansas coal, containing 15 to 20 percent volatile matter, well below Illinois coal's 35 to 45 percent. After receiving Interstate Commerce Commission permission, the railroad began delivering the coal, which could retail for as little as $5.50 per ton, much less than the cost of coke or other processed coals. In the end, the solution to the St. Louis smoke problem did not require better means of consuming Illinois coal. Rather, it required a means of acquiring better coal.[34]

The winter of 1940–41 quickly proved the value of the new approach. Stoker installations jumped, and despite some real hardships on the poorest coal consumers, smokeless fuels remained available through the cold months.

Early in the heating season, as the usual smogs developed across the river and beyond the reach of the new law in East St. Louis, the *Post-Dispatch* ran a cartoon of St. Louis asking out of mock jealousy, "Where's *My* Smoke?" Months later the U.S. Weather Bureau senior meteorologist noted that the winter passed "without a really good smoke," and comparisons made with the previous winter showed a 83.5 percent decrease in the hours of thick smoke. Subsequent winters proved the first was no fluke, and despite complaints from coal dealers the city continued to make progress under the new law.[35]

The impact of the St. Louis ordinance reached well beyond that city's limits. Its novel approach and early success drew broad attention, particularly from activists and officials in the smokiest cities, who quickly learned the value of shifting the regulatory focus from the boiler room to the coal yard. In Pittsburgh, the St. Louis law provided a model for further action. Like other industrial cities, Pittsburgh suffered deeply during the Depression, and policies that might threaten economic recovery gained little support. Indeed, excepting the W.P.A. survey conducted through the Health Department, the city accomplished little regarding smoke control during the Depression. Beginning in 1940, however, with the economy rapidly improving, and with the St. Louis model showing results, the city of Pittsburgh launched its own successful crusade.[36]

Pittsburgh experienced a number of extremely smoky days in the winter of 1940–41, and the *Press* reported extensively on both Pittsburgh's clouds and St. Louis's ordinance. As other newspapers began to support smoke control, residents lobbied for action. The city's Public Health director, Dr. I. Hope Alexander, traveled to St. Louis with several other Pittsburghers to investigate that city's progress. The trip inspired a similar junket by the city council undertaken so that officials might study the St. Louis ordinance at work.[37]

The renewed activism of early 1941 forced the mayor to appoint the Commission for the Elimination of Smoke. Among its members was H. Marie Dermitt, a founding member of the Smoke and Dust Abatement League in 1913, and now a member of the Civic Club's smoke elimination committee. Dermitt had familiarized herself with the St. Louis ordinance and supported its adoption in Pittsburgh. Other members of the commission, including Dr. Alexander and Councilman Abraham Wolk, shared her enthusiasm for the new model. Other commissioners, including A. K. Oliver of the Pittsburgh Coal Company and Patrick Fagan of the United Mine workers, had reason to be suspicious of any plan that might harm local markets for Pittsburgh coal.[38]

Sensing the seriousness of the threat to coal, the Western Pennsylvania

11. Where's My Smoke? At the beginning of the first heating season under a new smoke law, St. Louis looked with mock jealousy at East St. Louis, Illinois, which still lay under a pall of smoke. St. Louis *Post-Dispatch*, 19 November 1940.

Coal Operators Association, representing seventy-six operators in and around Pittsburgh, developed its own plan to reduce smoke in the city. Emphasizing the need to ensure that the region's coal would be a part of the solution, the association recommended a gradual approach involving the phasing in of processed coal or coke, the production of which would require new capital investment on the part of coal operators. The association, which made a report to the mayor's commission, made clear that the coal industry had an interest in finding a solution to smoke—the survival of its product depended on it.[39]

The coal industry's concerns apparently carried less weight than those of several other constituencies that testified before the mayor's commission. In

March, ten physicians implicated smoke in the city's high pneumonia death rate, as well as its colds and "sinus troubles." Several middle-class women also testified before the commission, as hundreds of others looked on in support. In addition to pressing the traditional arguments concerning cleanliness and health, one woman claimed that smoke had contributed to the exodus to the suburbs and the deterioration of the "Golden Triangle," Pittsburgh's downtown. The press followed this testimony closely, giving ample space to the complaints of the women and the pronouncements of the physicians.[40]

In short order, the commission recommended a St. Louis-type ordinance to the city council and the council passed it without major revisions in July, just five months after the mayor appointed the commission. As in St. Louis, the new law required all fuel consumers to install mechanical stokers or consume smokeless (low-volatile) fuels. Industries, offices, hotels, and apartments would be forced to comply by 1 October 1941, followed by railroads the next year, and domestic consumers in 1943. The ordinance passed with remarkably little debate; even Pittsburgh Coal's Oliver supported the law, as did representatives of the Steel Worker's and Miner's unions.[41]

As was often the case, passage of the new law proved easier than implementation. At the very outset the war delayed enforcement, with council pushing back compliance for domestic consumers to six months following the war's conclusion. But after the war, the coal industry continued to argue against implementation, noting that inadequate smokeless fuel supplies would unfairly burden the poor with high fuel costs. A postwar debate ensued among three interest groups: the coal industry, represented by the Western Pennsylvania Conference on Air Pollution, formed specifically for the purpose of delaying implementation; antismoke activists, now represented by the United Smoke Council, an umbrella organization created by the Civic Club; and redevelopment interests, led by Richard King Mellon's Allegheny Conference on Community Development. After agreement between the coal industry and redevelopment forces, and at the protest of many antismoke activists, the city finally passed new legislation requiring enforcement for domestic consumers on 1 October 1947, four years after the original implementation date.[42]

The first heating season under the new law did bring problems with low fuel supplies and high prices, but the city followed a policy of lax enforcement against the poor. Problems notwithstanding, the new ordinance brought immediate results, as the city saw considerably less smoke and more sun during the winter of 1947–48. By 1955 the Bureau of Smoke Prevention

12. Fifth Avenue before Smoke Control. In 1945, Pittsburgh remained a gloomy place at eleven in the morning. Courtesy of the Carnegie Library, Pittsburgh.

recorded only 10 hours of "heavy" smoke and 113 hours of "moderate" smoke, down from 1946's 298 "heavy" and 1,005 "moderate" hours. Clearing the smoke became an important step in the Pittsburgh Renaissance, and in redefining the city.[43]

The dramatic success in Pittsburgh cannot be attributed to the new ordinance alone. As in St. Louis, the supporters of the law anticipated that the city would turn to processed local coal for fuel. Instead, as in St. Louis, a new fuel source provided the key to rapid success. In this case, new pipelines from the Southwest provided a reliable and inexpensive supply of natural gas in the years following the war. By 1950, 66 percent of Pittsburgh households heated with natural gas, up from just 17 percent ten years earlier. As the more convenient gas gained favor in Pittsburgh homes, the most troublesome source of coal smoke disappeared. Even in this city where coal had so long played a crucial role in the economy, for miners and industrial workers, for marketers and hucksters, for businessmen and capital, homeowners willingly abandoned the troublesome fuel as soon as the supply of natural gas provided similar prices and reliability. Pittsburgh, like other cities around the nation, also witnessed the shift from steam locomotion to diesel-electric, a transition that occurred fairly rapidly in the years after World War II. By 1951, nearly half of the nation's locomotive stock was diesel powered. Although diesel locomotives are highly polluting in their own right, as this transition continued Pittsburgh and other cities benefitted from the decreasing pollution that it brought.[44]

13. Propaganda Smog. Downtown shoppers make the case for cleaning the air, 1946. Courtesy of the Carnegie Library, Pittsburgh.

14. Fruits of Victory. The Liberty Bridge and Tunnel in Pittsburgh, 1945 (*left*) and 1951 (*right*), four years after smoke abatement went into effect. Courtesy of the Carnegie Library, Pittsburgh.

Eventually all smoke-plagued cities would benefit from a shift toward natural gas heating and diesel locomotion, but none so dramatically as Pittsburgh. However, other cities did realize the importance of domestic smoke in the 1930s and 1940s. As early as 1928 Cincinnati's Smoke Abatement League began to refocus its attention, shifting from industrial power plants and railroads to residential hot-air furnaces. The Depression interceded, however, before the league could alter its procedures. With a 1933 budget at just 57 percent of 1928 expenditures, the league cut back on its staff, leaving just one of its three inspectors in the field. At the same time, domestic smoke worsened, as homeowners turned to less expensive and smokier fuels during the hard times. But the Depression did not necessitate the abandonment of smoke abatement efforts. The league conducted annual soot fall studies through the 1930s and indeed through the war. Not unlike federally funded investigations in other cities, these studies offered relatively reliable data concerning the sources of smoke. As Smoke Abatement League Supervisor Frank Lamping concluded in 1935, "The small plant operator presents the most immediate problem in the program. First, because they are so numerous, and; second, because of the tendency towards increased fuel prices."[45]

Residential fires posed entirely different problems for smoke abatement officials than did industrial boilers and locomotives. Officials could hardly expect homeowners to make extensive capital improvements solely to prevent smoke, nor was it reasonable to expect everyone who might work a domestic fire to become expert in combustion. Like those in other cities, Cincinnati officials quickly concluded that they could not use their experience in reducing industrial and commercial emissions as the basis of a residential smoke solution. As early as 1937, Cincinnati's city council debated a coal classification system, which would have required dealers to label coal by volatile content. In theory, this system would have allowed consumers to make educated decisions about their fuel and help ensure that householders could purchase the same grade of coal consistently.[46]

As in Pittsburgh, Cincinnati antismoke activists decided to follow the St. Louis example in 1940. With the support of the Smoke Department and the Smoke Abatement League, the city council's law committee began hearings on a St. Louis-type ordinance in March. By early April, the bituminous coal industry had launched a highly organized attack on the bill, which would have allowed the city to force consumers to change their fuels. Bituminous operations had much to lose if Cincinnati should restrict consumption. As the most accessible large city to the bituminous fields of West Virginia, eastern Kentucky, and southern Ohio, Cincinnati had become an important market for high-volatile coals, and an important transhipment point for coal-laden barges along the Ohio River.[47]

Actually, coal companies did so much business in Cincinnati that they could exert pressure on several local firms. The Blue Diamond Coal Company, for example, wrote to its suppliers in early April asking for assistance in quashing the bill. In a letter to Francis H. Leggett and Company, Vice President Fred Gore noted that "the Blue Diamond Coal Company purchases a substantial quantity of merchandise for its mines and stores from manufacturers and suppliers in Cincinnati." Gore then suggested that his firm would not want to take its business elsewhere. Heeding the not-so-veiled threat, Leggett's division manager followed Gore's instructions and sent letters condemning the smoke bill to city councilmen. One of those councilmen, Charles P. Taft,[48] responded with a hint of anger: "Especially because no copy of the ordinance has been furnished to councilmen, I am more than ready to listen to any reasonable arguments on the subject, but I do object to the campaign of misrepresentation which is being organized in the matter, from West Virginia."[49]

Taft received many other pleas to oppose the bill. Robert Castellini, who operated a prominent fruit and vegetable wholesale business, wrote, "We feel that this ordinance is excessively regulatory, dictatorial and discriminatory." In another letter, the Lower Cincinnati Business Men's Association, which represented several waterside wholesalers, concluded, "We do not believe the smoke situation, here in Cincinnati, is so serious that it could not be handled as in the past many years—through co-operation and education." Taft responded to this epistle with a bit of humor: "I think you will have to recognize . . . that the cooperation over a period of years has not produced the results that you and I want. At least I think if you consult your wife you will get that response."[50]

The coal companies' attack on the St. Louis-type bill became even more sophisticated in 1941, as the city council debated the issue anew. Several companies joined in founding the Coal Producers Committee for Smoke Abatement, which would operate out of Cincinnati through the early 1950s. During those years, the committee struggled to keep southern and midwestern markets open to high-volatile coals by guiding antismoke efforts away from fuel-restricting ordinances. By 1949, the committee represented twelve high-volatile coal operators associations and seven coal-bearing railroads. The committee conducted smoke surveys in several cities, offered consultation for polluting plants, prepared sample ordinances for municipalities, distributed information on the smokeless firing of high-volatile coals, and engaged in a public relations campaign in support of the fuel. By 1949, the committee had already successfully worked to keep high-volatile coals in Cleveland, Philadelphia, Providence, Winston-Salem, Durham, Charlotte, and Greenville. It had failed, however, in several important cities, including Cincinnati, Pittsburgh, and Milwaukee.[51]

In Cincinnati the committee operated through the Metropolitan Smoke Control Committee, which it helped organize. This committee arranged for the testimony of several coal men before the law committee of the city council, including Howard Eavenson, the president of Bituminous Coal Research, an organization formed by the National Coal Association to combat the continuing diminution of coal markets. The director of the Cincinnati Coal Exchange also testified, as did representatives from Appalachian Coals, Inc., and Jack Vogele of the Coal Producers Committee for Smoke Abatement. Together these men warned that an ordinance restricting high-volatile coals would work great hardship on lower-class city residents and could cause severe fuel shortages. Vogele disputed claims about the effectiveness of the St. Louis ordinance, arguing that smoke had decreased significantly during the first heating season under the law because of the mild winter, not because of coal restrictions. He also produced newspaper clippings reporting continuing smoke problems in St. Louis and the fining of coal dealers who had failed to obey the new ordinance.[52]

Even while arguing against the St. Louis-type bill, the coal men offered their own smoke abatement ordinance. Although this bill did propose the labeling of fuel by volatile content, it did not restrict fuel consumption or require the licensing of fuel dealers and truckers. Remarking on the advantages of this bill, Julian Tobey, vice president of Appalachian Coals, claimed that it "visits no hardships on persons in the low income brackets," and that it "does NOT penalize coal consumers by forcing them to buy high priced fuels, even though they have created no smoke nuisance."[53]

The Smoke Abatement League fought back, bringing Tucker himself to testify before the law committee, where he gave a more positive assessment of the progress in St. Louis. Dr. Louis T. More, long of the league, testified as well, as did league supervisor Frank Lamping. Also in favor of the bill was Dr. Clarence Mills, a physician at the University of Cincinnati who would dedicate his career to defining the relationship between smoke and human health. Mills had already begun to connect air pollution with lung cancer and other respiratory diseases through his studies comparing death rates in cities and the suburbs.[54]

Although the war distracted public attention from Cincinnati's smoke problem, city council passed a new control law in September of 1942. A compromise ordinance, it did not limit coal use within the city, but it did expand the Smoke Department and force dealers to label their coals by volatile content. In practice the new law made little difference, as the war slowed the expansion of the department, and indeed meant the loss of Chief Inspector Charles Gruber, who entered the service in 1941 and was not immediately replaced. Dealers also thwarted coal labeling objectives by labeling most their coals high enough to prevent consumers from distinguishing them by labeled content.[55]

Bituminous coal interests could hardly have been more pleased with the new ordinance, except perhaps if it had worked sufficiently well to prevent the passage of a new ordinance shortly after the war. Indeed, association of the failed law with coal interests may have sped the passage of its replacement. Referring to the 1942 law as "the coal men's ordinance," Taft exclaimed, "What a dirty record it has!" Taft and his fellow Charterite[56] councilmen again supported a new St. Louis-type bill in 1946, this time successfully. By the spring of 1947, Cincinnati had restricted consumption of high-volatile coal. As in St. Louis and Pittsburgh, the new ordinance precipitated significant improvements in air quality. Studies conducted by the Smoke Abatement League indicated that soot fall dropped 50 percent in the first eight years under the law.[57]

Although Pittsburgh and Cincinnati found success in restricting high-volatile coal, not every smoke-plagued city followed the St. Louis model so directly or so quickly. New York, which had always differed from the midwestern metropoli in both the density of its smoke and in its attempted solutions, did little to combat its worsening problem until after World War II. Through the 1930s and early 1940s, the Health Department retained jurisdiction over smoke control, and it dedicated little manpower to the problem. But when the crises of depression and war had passed, activists once again drew the city's attention to the growing smoke clouds.[58]

The New York *Times* played a key role in the renewed campaign, running dozens of smoke-related editorials and letters to the editor, particularly after the first of the year in 1947. Noting Health Department figures that showed a 40 percent increase in soot between 1936 and 1945, the *Times* announced, "We have a serious smoke problem in New York, as many a housewife, appalled by the soot damage, will testify." The *Times* ran a campaign which would not have seemed out of place in the 1890s, with an emphasis on health and cleanliness, and even at times, morality. One letter from a reader in Glens Falls declared, "Smoke is not only bad for health, property and morals but, I believe, is a contributing factor in crime and corruption." Apparently smoke could still represent the worst qualities of cities to many Americans. The *Times* even implicated smoke and soot in the suburban migration then under way.[59]

The movement quickly gained the support of Mayor William O'Dwyer and Joseph T. Sharkey, the Democrat's majority leader in the city council, but even as the *Times* reported on the great strides taken by Cincinnati and Pittsburgh, New York was slow to create a plan of attack. Once praised for its remarkably clean air, New York was rapidly becoming one of the most polluted cities in the country, as it moved through a fuel transition. Anthracite continued to lose market share in the city, and by 1948 bituminous coal outsold its cleaner rival. Simultaneously, fuel oil made significant inroads in the domestic heating market. Combined with ever growing amounts of automobile exhaust, emissions from oil and bituminous fires contributed to a rapidly deteriorating air quality.[60]

Two interest groups joined the movement in 1948. The Citizen's Union set up its own air pollution commission, chaired by Arthur Stern, the chief engineer at the State Department of Labor's Division of Industrial Hygiene. Stern had considerable experience with the issue, including heading the W.P.A. air pollution survey in the city a decade earlier. Stern aided councilman Sharkey in drafting another bill, one that would move control of air pollution from the Health Department to the Housing and Buildings Department, which already had the authority to inspect fuel-burning equipment. The other interest group, the Outdoor Cleanliness Association, offered less direct but equally important support for government action. With a membership list that included some of the city's wealthiest and most powerful families, many of them residing in the Upper East Side, the association focused considerable attention on the issue. The *Times*, for example, regularly reported the association's actions, perhaps because one of its members, Iphigene Ochs Sulzberger, was the wife of Arthur H. Sulzberger, the paper's publisher.[61]

As the city inched toward legislation, with the Sharkey bill receiving a full hearing in the council, events in Pennsylvania made front-page news. On 31 October the *Times* ran a short article under the headline, "'Smog' Linked to 18 Deaths in Day and Hospital Jam in Donora, Pa." The next day a fuller report put the death toll at twenty, noting that many residents had left the area and that volunteer firemen were delivering oxygen to those who stayed behind. The article blamed the American Steel and Wire Company's zinc smelter for the disaster, quoting Norbert Hochman, a chemist with Pittsburgh's Smoke Prevention Bureau, who speculated that sulphur trioxide from the plant had provided the toxicity. On the same day the *Times* also ran an article reminding its readers of a similar fate which befell Belgian residents of the Meuse valley in 1930, when a sulphurous smog was blamed for seventy deaths, particularly among sufferers of asthma and heart disease. New York's Health Department quickly assured city residents that no such disaster could strike the metropolis, but as the *Times* reported, Ohio's governor was taking no chances. Just two days after the disaster, he requested that the state health director report on atmospheric pollution in Ohio's industrial areas as a first step in preventing a similar event there. Three weeks after Donora's killing smog, Dr. Clarence Mills, now the head of the University of Cincinnati's department of experimental medicine, reported on his trip to the city. Mills related the patterns of acute reaction to the smog, especially the gender imbalance among the dead (fifteen men, five women), to his observations of chronic disease and air pollution. Smog, Mills concluded, was indeed a health threat in many communities.[62]

Despite the attention, the deaths in Donora did not convince New York City to move more swiftly toward a solution for its own problem. The new bill could not pass council until February the next year, and not until July did the mayor appoint a director for the newly created Board of Smoke Control. The board removed the pressure from an understaffed Health Department, and placed air pollution control in the hands of an understaffed engineering bureau. As complaints continued to pour into the new office, the director, William Christy, an engineer and a long-time member of the Smoke Prevention Association, counseled patience. Christy and his staff issued a litany of warnings and instructions for violators, an approach not unlike that taken by the city for over forty years.[63]

Many New Yorkers could no longer oblige the go-slow approach. As "An Irate Housewife from Gramercy Park" wrote, "If Pittsburgh and St. Louis could clean up their cities surely New York could do the same." By mid-1950, concern over the failure of the 1948 ordinance had created yet another movement to change the city's law. That summer Mayor O'Dwyer invited Raymond Tucker to visit New York and offer suggestions. Not surprisingly, Tucker recommended fuel guidelines similar to those in St. Louis. Out of Tucker's suggestions came a bill allowing the Smoke Board to set limits on volatile content in coal and sulphur content in fuel oil. As in other cities, this proposal garnered considerable opposition, particularly from the fuel industry. In a public hearing of the bill, Herbert Lammers of the Coal Producers Committee for Smoke Abatement labeled the bill a "fuel abatement rather than an air pollution abatement document." Much to the dismay of the coal industry, however, two women from the Yorkville neighborhood of Manhattan won the day when they pulled dirty laundry from a paper bag to show the hearing's crowd of four hundred. After displaying a soiled sheet and shirt, Mrs. Daniel Dolan asked simply, "How much longer are we going to have to wait for results on smoke control?" The *Times* lauded the women's performance, claiming that their simple presentation and the applause that followed all but eclipsed the more technical arguments of the coal men.[64]

When the new bill became law later that year it had been stripped of its restrictions on the sulphur content of fuel, but provisions limiting the volatile matter in coal (to 24 percent) remained intact. The new law also took steps to improve historically lax enforcement, particularly by increasing the number of city smoke inspectors. The coal industry had good reason to protest the new law, since their products faced greater restrictions than did coal's chief competitor in New York, fuel oil, which would have faced restrictions had the sulphur clause passed.[65]

Despite the public agitation for control which had sparked the passage of the new law, the Smoke Board did not rigorously enforce the 1950 ordinance. Indeed, as of January 1951 the city had not issued a court summons for a smoke violation since the passage of the 1949 ordinance. In the face of this inaction, Elizabeth Robinson, a leader in the New York Federation of Women's Clubs, organized a new interest group, the Committee for Smoke Control. To determine the effectiveness of the new law, Robinson surveyed housewives on the issue: 62 percent thought that the city's air was dirtier than previous years, 32 percent reported no improvement, and just 6 percent thought the city was cleaner under the new law. Although unscientific, Robinson's survey indicated that the city had not convinced middle-class women that progress had been made. It also indicated that the city needed a reasonable means of determining air quality.[66]

The city did begin a "get tough" policy, which led to several prosecutions, but Robinson kept the pressure on. In May 1951, the Committee for Smoke Control distributed thousands of pamphlets on air pollution, focusing on health effects and the failings of the smoke bureau. The British-born Robinson, a housewife from Forest Hills, Queens, may have been motivated by her own childhood struggles with bronchitis, which she blamed on Liverpool's smoke. "Women are thinking of the health of their children," she told the *Times*, "and where that is at stake, a woman will fight as primitively as any tiger in a jungle."[67]

With the Smoke Bureau muddling along, without achieving the results that had greeted other cities using the St. Louis model, yet another movement for an improved ordinance developed. This time Mayor Vincent Impellitteri sent a bill to the council drafted by City Construction Coordinator Robert Moses. When the new law passed in September 1952, it created a new Department of Air Pollution Control. Two months later, when the law became effective, Impellitteri appointed Dr. Leonard Greenburg to head the new department. Greenburg, formerly the executive director of the Division of Industrial Hygiene and Safety Standards at the state Department of Labor, held an M.D. and a Ph.D. in public health. New York's new department and director marked a fresh direction in municipal air quality control. This new approach not only put the protection of human health at the center of its mandate, but also recognized the complexity of the modern air pollution problem. No longer could "smoke" suffice as the description of atmospheric pollution, particularly with greater contributions coming from automobiles, buses, incinerators, and oil-burning industries. These changes in atmospheric pollution accompanied the fuel transition then well under way in New York and the rest of the nation.[68]

Other cities made similar changes, including Cleveland. In 1947, after an extensive campaign by the Coal Producers Committee for Smoke Abatement to prevent the adoption of a St. Louis-style ordinance, the city chose another route. Instead of revising its regulations, Cleveland reorganized its regulators. The 1947 law created the Division of Air Pollution Control, to reside in the Department of Public Health and Welfare. This new division gathered the Bureaus of Industrial Hygiene, Industrial Nuisances, and Smoke Abatement, and for the first time the city addressed all air pollution, both indoor and outdoor, from one office. This reorganization reflected a new understanding of the problem, one that recognized the great diversity of air pollution hazards, particularly from industrial chemicals that posed threats to workers and neighbors alike. The Smoke Bureau, then, moved from the Department of Public Safety, to join other pollution-control offices in the Health Department, a movement repeated in many other locations, including the federal government, where air pollution responsibilities moved out of the Bureau of Mines and into the Public Health Service.[69]

This shift reflected the coming of age of industrial hygiene as a recognized profession. As historian Christopher Sellers has argued, by the 1940s experts in work-related disease had begun to look outside the factory walls, toward consumers and the broader environment. This environmental turn of industrial hygienists became particularly clear after the Donora disaster in 1948, when the Public Health Service's Division of Industrial Hygiene led the government's investigation. As the pioneering Dr. Robert Kehoe noted in 1958, "The industrial environment has become, to a remarkable extent, the national environment." The expanding authority of industrial hygienists came at the expense of power engineers. No longer would air pollution expertise find its grounding in steam engineering. Henceforth, chemists and M.D.s would dominate the environmental health sciences, as the credentials of industrial hygiene eclipsed those of industrial engineering; henceforth, debates over air pollution would focus on the determination of safe concentration levels of any number of pollutants, rather than on the development of more efficient means of coal combustion.[70]

The reorganization of Cleveland's air pollution bureaucracy also reflected the decreasing importance of coal smoke in the overall composition of air pollution, which was partly a consequence of the decreasing reliance on coal in Cleveland, and many other cities. In its first report, the Bureau of Smoke Abatement reported that in 1946 "the amount of solid fuel which was replaced by oil and gas was estimated to be between 375,000 and 400,000 tons." By 1950, Cleveland, like Cincinnati and Pittsburgh, would be more reliant on natural gas than coal for domestic heating. For the coal industry, then, the success in preventing restrictive legislation in Cleveland proved hollow, as that city, like others, abandoned coal for more convenient fuels.[71]

Although antismoke efforts had taken a new tack, in one respect the movement had come full circle. With more and more research verifying that chronic exposure to smoke caused serious health problems, and with deaths attributed to acute air pollution events, particularly in Donora, the smoke issue once again became primarily a health issue. The revival of health concerns became even more apparent as the nation began to struggle with the more complex air pollution problems related to petroleum combustion. The return of municipal smoke-control authority to health officers, physicians, and chemists highlighted the renewed importance of health and the diminishing role of the steam engineers who had dominated smoke abatement efforts for thirty years.

conclusion

The Struggle for Civilized Air

From the 1910s through the 1930s, engineering smoke experts had predicted that scientifically operated municipal departments issuing education and advice would solve the problem of smoke pollution—even if they did so gradually. The experts were mistaken. Real relief came only with significant changes in fuel use, as urban America became less dependent on soft coal in the mid-twentieth century. For decades visionaries had predicted this fuel transition and promised the advent of cleaner energy technologies, even as the nation clung to its cheap, dirty coal. As early as 1896, Francis Crocker envisioned "coalless cities" in *Cassier's Magazine*. Electricity, Crocker argued, would make cities smokeless and thus healthier. In 1908, as Annie Sergel and other women heightened the antismoke campaign in Chicago, George Weber announced that the city's smoke problem was gradually solving itself. Weber argued that increasing use of natural gas and electricity, combined with better construction of all future coal-fired steam plants, eventually would bring an end to coal smoke in Chicago. Of course, the predictions of Crocker and Weber did come true, but undoubtedly both would have been surprised at how long the transition from dirty coal to clean gas and electric technologies actually took. As the Bureau of Mines's O. P. Hood remarked in 1930, after nearly twenty years of personal involvement in the antismoke effort, time had proven that smokeless fuels had to be as inexpensive as coal to gain wide use.[1]

Not until after World War I did cleaner sources of energy begin to meet the nation's increased demand for power, and thereby curtail the growth of coal consumption. Before the war, petroleum products did become important in fueling transportation, replacing coal on steamships and warships and, of course, in propelling an expanding fleet of automobiles. But as an urban fuel, petroleum only very slowly replaced coal, with most prewar use coming in the cities of the West and Southwest, including Los Angeles, San Francisco, Houston, and Dallas, where abundant oil was cheaper than coal, which had to travel great distances to reach these growing energy markets. The availability of oil kept these flourishing cities relatively clean, and in later decades an expanding oil supply and improved delivery technology would help reduce other cities' coal dependency. Although many New Yorkers predicted oil would replace coal much earlier, particularly during the 1902 anthracite strike, oil made significant inroads in the northern fuel market only after World War II.[2]

As urbanites would increasingly become aware in the 1940s and 1950s, petroleum fuels did not provide the cleanest of alternatives to coal, nor did oil offer the greatest hope for relief from coal smoke. Natural gas, which had rendered relief to Pittsburgh in the 1880s, continued to gain wide support from antismoke activists. Indeed, natural gas interests used the fuel's smokelessness as one of its central selling points. Before the war, however, limited supplies and distribution networks meant a limited role for gas in the fuel market. In the 1920s, new pipelines greatly reduced the price of natural gas in the South and the West, with significant increases in natural gas consumption occurring in California, Texas, Oklahoma, and Louisiana; but expansion in the smoky Midwest remained limited, particularly in industry. Not until the 1940s and 1950s did natural gas come to most midwestern and eastern cities in sufficient quantities to compete with coal. Transportation costs remained high for natural gas, as they did with all fossil fuels. These high costs explain the creation of regional fuel markets, with oil, gas, and coal remaining cheapest and most widely used near their sources.[3] (See tables 2 and 3.)

Although natural gas and oil provided cleaner alternatives to coal, the switch to electricity offered the most significant opportunity for movement to cleaner energy technology. The production of electricity did require massive consumption of some fuel, which in cities suffering from smoke problems was generally coal, but centralized electric power plants consumed fuel more efficiently than the dispersed steam boilers common early in the century. A single central electric power plant could displace thousands of smoking stacks, in a sense creating a "point source" for smoke emissions. Central power plants received intense scrutiny from engineers seeking efficiencies and from regulators seeking compliance. This meant that the centralization of power creation allowed for a dramatic reduction in emissions per btu, as power companies continued to innovate to squeeze more energy and less smoke out of every ton of coal. Just as important, manufacturers used the flexibility of electricity to make their businesses more energy efficient.[4]

Like natural gas and oil, electricity had long been touted as a savior of urban atmospheres. In 1901, as the U.S. Geological Survey prepared experiments to burn coal smokelessly at the St. Louis Exposition, the New York *Tribune* announced that electricity, not smokeless coal firing in boiler plants,

TABLE 2
The Relative Decline of Coal: Energy Sources as Percentages of Total Energy Consumption

Year	Bituminous coal	Anthracite coal	Oil	Natural gas
1890	41.4	16.5	2.2	3.7
1900	56.6	14.7	2.4	2.6
1910	64.3	12.4	6.1	3.3
1920	62.3	10.2	12.3	3.8
1930	50.3	7.3	23.8	8.1
1940	44.7	4.9	29.6	10.6
1950	33.9	2.9	36.2	17.0

NOTES: Measured in Btus
   SOURCE: Adapted from Sam H. Schurr and Bruce C. Netschert, *Energy in the American Economy, 1850–1975* (Baltimore: Johns Hopkins Press, 1960), 36.

would eventually free cities from smoke. The *Tribune* envisioned large central power plants in cheap suburban lands or even along distant waterways as the ultimate in smokeless energy suppliers, a prediction that has largely come to pass.[5] While electricity did gain widespread use before the World War I, not until after the energy crisis of the war years did most industries begin the shift to electric power. As historian Harold Platt has argued, the shock of the energy crisis convinced many energy consumers to remove themselves from the volatile fuel markets and to rely instead on electric power companies for energy. In the 1920s, the switch to electric power in industry marked a second industrial revolution, as the remarkably flexible energy source sparked the development of a myriad of new machines, for use at home and in the factory. Of course, electricity did not so much compete with coal as provide it with a new and growing market. While hydroelectric plants did produce up to one-third of the nation's electricity through the 1930s and 1940s, much of the thermal generation of electricity came from coal. By 1955, the nation burned 140 million tons of coal to produce electricity. This meant that nearly a third of the nation's coal production went into the generation of electricity, a percentage that continued to rise.[6]

   The highly competitive nature of the soft coal industry contributed to labor unrest and dramatic price fluctuations, particularly in the 1920s and 1930s. Chaos in the bituminous industry made alternative fuels more attractive, but the transition away from coal-fired boiler technology was a long one. The gradual nature of this energy transition meant the persistence of smoke problems in most industrial cities into the 1940s and early 1950s. Still, the general stagnation of coal consumption from 1920 through the 1950s, despite the increasing demand for coal in electricity generation, revealed the depths

TABLE 3
Value of Fuels as Percentage of Total Value of Fuels Consumed

Year	Bituminous coal	Anthracite coal	Oil	Natural gas
1890	51.6	30.5	9.9	8.0
1900	59.4	22.9	13.0	4.7
1910	61.3	20.8	14.3	3.6
1920	49.4	10.8	36.1	1.9
1930	30.4	13.6	45.4	5.6
1940	32.7	7.8	52.0	4.6
1950	24.9	4.0	62.1	4.3

NOTES: Based on average unit prices
SOURCE: Adapted from Sam H. Schurr and Bruce C. Netschert, *Energy in the American Economy, 1850–1975* (Baltimore: Johns Hopkins Press, 1960), 536–37.

of the industry's problems. Through these decades coal remained abundant (often overabundant) and cheap, coal consumption technology remained dependable and ubiquitous, and the steam engineers who had helped build a coal-dependent nation remained an important force within industry and the nation's engineering schools. How, then, did coal lose so much market share to its chief competitors, natural gas and oil?

The answer lies largely in the economic advantages of the competing fuels themselves. Where available, natural gas was even less expensive than coal, and while oil was more expensive as a fuel in the Midwest and East, its use could reduce operating costs. Railroads, for example, had accounted for 25 percent of the nation's coal consumption in 1920, but by 1955 they consumed less than 4 percent of the black fuel. The rapid transition of the nation's locomotive stock from steam to diesel after the Second World War followed a general acknowledgment within the industry that diesel engines provided significant operating economies, justifying the capital costs involved in the transition. The shift from steam locomotion to diesel constituted a serious setback for the coal industry, but cities benefitted greatly from the disappearance of one of their most troublesome smoke sources.[7]

Coal also clearly lost market share for noneconomic reasons, including its smoky nature. As Herbert Lammers of the Coal Producers Committee for Smoke Abatement argued in 1949, "Smoke, fly ash and dirt is not the only resistance that our coals have to meet in dealing with the public, but it is one of the principal ones today, perhaps the outstanding one, that is turning away users. The public of today insists upon cleanliness and will have it." In a talk before the Virginia Coal Operators Association, Lammers conceded, "You can't really blame the housewife for agitating for clean air. She is used to the modern kitchen, the mechanical icebox, the vacuum sweeper, the automatic furnace and water heater. She wants clean air to go along with these modern improvements." With this Lammers nicely narrated what historian Samuel Hays has identified as the root of postwar environmentalism: the middle-class demand for environmental amenities to accompany the many other benefits of affluent life. Lammers understood that modern sensibilities valu-ing convenience and cleanliness also devalued coal.[8]

Lammers realized that King Coal was surely dead, and that the industry would have to battle vigorously to maintain market share. In 1950 coal provided less that 30 percent of the country's total btu's; just forty years ear-lier it had supplied over 80 percent of the nation's energy needs. While the industry still held economic and political influence in the major coal-producing states, it had been so weakened by competition and antismoke agitation that cities could successfully legislate against the sale of its prod-ucts—an unthinkable proposition just three decades earlier. Indeed, writing as late as 1933, smoke abatement proponent and Consolidated Gas Compa-ny executive Henry Obermeyer conceded, "We cannot conceive of a law so drastic that it would attempt to regulate the type of fuel to be burned." Just six years later such a law had passed, and others were on the way. Although these laws affected only high-volatile coals, Lammers warned in 1949, "Re-member if it is legal to bar one kind of coal, it can become legal to bar all coals." This was not a caveat spoken from a position of confidence.[9]

Struggle as Lammers and the Coal Producers Committee for Smoke Abatement might against the energy transition, coal's loss of market share in certain sectors was permanent, and eventually nearly complete. Never again would coal provide significant amounts of energy for the transportation sec-tor, and coal would continue to decline in importance in the domestic heat-ing market (except in its role in producing electricity). Coal also suffered loss-es as a material source for the chemical industry, which increasingly turned to petroleum and natural gas for its needs.[10] (See table 4.) These shifts away from coal, combined with the ever-increasing consumption of petroleum products, precipitated an air pollution transition. Even as cities began to im-plement plans to control coal smoke through fuel restrictions, other sources of air pollution gained in importance. Indeed, expert and public concern for "smoke" itself gradually shifted to a concern for "air pollution," as the com-plexity of the problem became more evident.

As city governments created new air pollution control bureaus, which subsumed older smoke bureaus, so too did the private sector redirect its fo-cus to match the changing reality. In 1950, the Smoke Prevention Association recognized the transition under way and changed its name to the Air Pollu-

TABLE 4
*The Stagnation of Bituminous
Consumption: Apparent Consumption*

Year	Bituminous coal
1890	110,785,000
1900	207,275,000
1910	406,633,000
1920	508,595,000
1930	454,990,000
1940	430,910,000
1950	454,202,000

NOTE: Rounded to thousands of net tons
SOURCE: Adapted from Sam H. Schurr and Bruce
C. Netschert, *Energy in the American Economy,
1850–1975* (Baltimore: Johns Hopkins Press, 1960),
508–9.

tion and Smoke Prevention Association, and just two years later it became simply the Air Pollution Control Association. A few years later, the venerable Smoke Abatement League in Cincinnati changed its name to the Air Pollution Control League. The press, too, gradually shifted its concern from smoke to the broader air pollution problem. By 1960, the *Readers Guide to Periodical Literature* abandoned "Smoke Prevention" as a heading, replacing it with "smog" and "air pollution," where most references to the topic had resided for more than a decade. One organization clung to its old name, however. Through the early 1950s, the Coal Producers Committee for Smoke Abatement, perhaps pleased that coal no longer deserved near-exclusive blame for the nation's air pollution problem, remained committed to *smoke* control and to publicizing the notion that oil and gas consumption also caused pollution.[11]

The changing terminology surrounding atmospheric pollution reflected the real and relative decline of coal emissions in the smog above growing numbers of cities. Two air pollution stories from the late 1940s played important roles in shifting public concern away from coal smoke and toward other sources of pollution. First, the Donora episode, in which industrial fumes, not coal smoke, were responsible for the deadly cloud, convinced many observers that nonvisible emissions were potentially much more harmful than coal's particulate emissions. All antismoke efforts and governmental responses had been predicated on controlling *visible* emissions, as evidenced by the use of the Ringlemann Chart. Once air pollution experts and activists rejected the notion that invisibility meant success, combating "smoke" through the use of a color scale made little sense. After Donora, more and more people would demand to know exactly what was coming from stacks—including visible and invisible elements.

As the nation's first acute air pollution attack with immediately attributable deaths, Donora also forever linked severe air pollution with health. Additionally, it changed the way many Americans perceived the problem. Writing in *Woman's Home Companion* in 1949, Clive Howard revealed the depths of these changes. In an article titled "Smoke: The Silent Murderer," Howard wrote at length about Donora, but also about "nauseating gases" from oil refineries around San Francisco Bay, and of a "sickening odor of oil or gas" in New Orleans which sent people running to vomit. Although he also discussed Clarence Mills's studies linking smoke with cancer, Howard did not refer to any instance where coal smoke caused an acute episode with immediate health consequences. Indeed, although his title accused "smoke," Howard mostly wrote about chemical fumes. Even Mills, who had studied smoke for years, concluded in 1954, "Today our interest has shifted from the dense black smoke and dirty smogs of yesteryear over to the grayish or blue-gray chemical smogs of the present." Even as health became the major concern for those combating atmospheric pollution, smoke itself became less of a consideration.[12]

The second story involved Los Angeles's regular summer blankets of smog, which obviously had little to do with coal since the city consumed only small amounts of that fuel. The peculiarity and severity of Los Angeles's pollution after World War II helped shift the focus of the nation's air pollution dialogue from the smoky Midwest to the smoggy Far West. L.A. had grown rapidly during the war, in both population and industrial production, and its citizens suffered under dramatically worsening air quality. In 1942 and 1943 the city experienced unexplained eye-stinging smog events, and the municipal and county governments struggled to find a solution (or even the cause). L.A. suffered not from dense, carbonous smoke clouds but from acrid clouds of complex chemical composition. In 1943 the city health officer identified fourteen different irritants present in the city's eye-stinging smog, including ammonia, formaldehyde, acrolein, acetic acid, sulfuric acid, hydrochloric acid, and nitric acid. Activists speaking out against the smog blamed a number of sources for causing the clouds, including automobiles, diesel busses, backyard incinerators, oil refineries, and other industries.[13]

Hoping to move the city toward a solution, in 1946 the Los Angeles *Times* hired Raymond Tucker to study the problem and make recommendations, just as he had for many other cities. In Los Angeles, however, Tucker was out of his element. With little coal consumption in the basin, Tucker could not simply recommend that the city require better grades of fuel. Similarly, with the wide variety of emissions, the actual source of the problem was difficult to discern. In his report, which the *Times* published in full, Tucker made many recommendations, including the banning of back-yard incinerators, extinguishing fires at municipal and commercial dumps, the monitoring of (largely oil-fed) industrial plants, and fining truck drivers who allowed their diesel engines to smoke. Tucker's only fuel recommendation involved the switching of railroad engines from oil to diesel. Given the fuel use patterns in the city, Tucker found no need to mention coal at all in his report.[14]

In 1950, Arie Haggan-Smit identified the major components of L.A.'s smog to be ozone, created through photochemical reaction of hydrocarbon exhaust, and unburned gasoline. Shortly thereafter the city would issue "smog alerts" based on the ozone content of the atmosphere. In an economy based increasingly on internal combustion engines, oil refineries, plastics manufacturing, synthetic rubber plants, and diesel locomotives, coal emissions meant very little. This was the case even outside of Los Angeles, as the entire nation's economy went through its uneven shift toward petroleum dependency, and its concerns for air pollution moved away from carbonous particles and toward invisible gases, such as ozone and unburned gasoline.[15]

In its new incarnation as the nation's most important source of electricity, coal would eventually come under fire as a major contributor to acid rain and global warming, but after 1960 almost no one would complain about coal smoke in American cities. The St. Louis, Pittsburgh, and Cincinnati movements of the 1940s notwithstanding, this near-complete smoke elimination came largely at the hands of a fuel transition—particularly the great diminishment of coal consumption in locomotives, homes, apartments, office buildings, and many industries. This suggests that smoke abatement efforts from the progressive era through the 1940s had little effect. Certainly the slow progress in smoke reduction before World War I, the utter failure to control smoke during that war, and the continuing problems into the 1950s all suggest that the antismoke movement of the progressive decades had little real influence on the urban environment. Indeed, several historians have reviewed the smoke abatement movement and judged the effort an unmitigated failure.[16] Still, while the lingering smoke evidenced real shortcomings, in many ways the movement's accomplishments were remarkable. In an industrializing, urbanizing economy dependent on coal, the antismoke movement struggled against the most powerful forces at work in the nation: economic and demographic growth. That the smoke problem did not increase dramatically in the progressive decades should suggest some limited success for the movement.[17]

The nation's industrial cities experienced exponential growth during the four decades surrounding the turn of the century. The 1880 census reported the combined populations of the cities most cited in this study, New York, Chicago, St. Louis, Pittsburgh, Cincinnati, Cleveland, and Milwaukee, at 3.3 million.[18] By 1920 nearly 11.3 million souls inhabited those same cities, an increase of over 340 percent. In total, by 1920 over 54 million Americans lived in cities, more than twice as many as in 1890, and by 1950 nearly 89 million people lived in America's cities. Most of the growth in urban population came in the nation's largest cities. The 1890 census reported only 9.7 million Americans lived in cities with over 100,000 inhabitants, but by 1920 almost three times as many Americans, over 27 million people, lived in cities that large. By 1950 that number had increased to nearly 43 million.[19]

In the same years, the national economy became ever more energy intensive. Per capita energy consumption (excluding wood) more than doubled from 1900 to 1950. More important, before 1920 aggregate coal consumption outpaced urban growth, with most of that growth coming from an expanding bituminous coal supply. Thus, as the antismoke movement reached a peak in activity just before American involvement in World War I, the nation was burning more and more dirty coal every year. In 1890, before antismoke movements began to organize, the nation consumed less than 111 million tons of bituminous coal. In 1913, when most industrial cities had created professional smoke inspection departments, the nation consumed nearly 460 million tons. In 1918, with the nation at war and antismoke laws ineffectual, the nation consumed over 556 million tons of soft coal, over five times as much dirty coal as the nation consumed when antismoke activism began.[20]

Thus, with larger and larger American cities burning more and more coal, the progressive-era smoke abatement movement may have done well to simply prevent the development of a greater environmental disaster. Just as important, the movement forced local and often state governments to assume regulatory authority over urban atmospheric emissions. This regulation marked an important first step in the public control of the industrial environment. Through the efforts of progressive-era reformers, the municipal authority to regulate emissions withstood judicial review, and the state asserted that the public right to clean air superseded individual rights to inexpensive atmospheric waste disposal in the pursuit of profit. Nevertheless, that municipalities regularly failed to defend the rights of the public over the rights of individual polluters indicated that victory would ring hollow for decades to come for many urban residents. Still, these early movements provided the crucial legal foundations for the more successful movements of the 1940s and 1950s.

The antismoke movement saw two transitions. In the 1910s, the lay reformers who had defined the problem left finding the solution in the hands of engineering experts. In doing so they also largely allowed the environmentalist antismoke movement to develop into a conservationist effort concerned with efficiency and economy. In other words, the movement evolved from a lay effort focused on smokestacks, to an expert-led endeavor focused on the boiler room. In the late 1930s, Raymond Tucker shifted the focus yet again, to the coal yard, where government officials and the public could readily understand the simple truism: bad fuel led to bad air. Tucker's innovation allowed a new generation of reformers to rebuild the old argument that a healthful, beautiful, and moral environment required clean air, and that society had to act to abate pollution, even at great expense. Concerns about health particularly regained a central place in the antismoke dialogue, and then played a pivotal role in efforts to control other kinds of air pollution.

Few cities had time to celebrate the end of coal smoke, whether it came at the hands of fuel-restricting legislation or by more gradual shifts in fuel use. Actually, the passing of coal smoke from city skies went largely unreported, as new types of air pollution created new problems for urbanites—problems that would take decades to solve. As O. P. Hood noted in 1930, at some point the public would raise its standards for its environment and demand cleaner surroundings. It would even come to demand that steps be taken to improve the environment before experts could complete all the research, before economists could quantify the discomfort of city residents and the unhealthfulness of individual pollutants, and properly juxtapose them with the costs of reform. At some point the public would demand action toward cleaning the air and reject calls for further, delaying study. Decades after the nation had forgotten the work of Pittsburgh's Women's Health Protective League, Cincinnati's Charles Reed, and Chicago's Annie Sergel, urbanites would again demand civilized air for their civilized nation.[21]

Notes

introduction Vision Obscured

1. "When the Milk Is Spilled," New York *Tribune*, 11 May 1899. These environmental disasters might be included in what historian Martin Melosi has called an "environmental crisis" in the late 1800s. Although Melosi focuses on urban problems—sewage, water pollution, smoke, garbage, and noise—the destruction of open space and the demolition of the nation's forests were also directly related to the growth of industrial cities. See Martin Melosi, "Environmental Crisis in the City," in *Pollution and Reform in American Cities, 1870–1930*, ed. Melosi (Austin: University of Texas Press, 1980), 3–31. The New York *Tribune* had long expressed interest in environmental issues, having supplied leadership in the effort to preserve the Adirondack forests in the early 1880s. See Roderick Nash, *Wilderness and the American Mind* (New Haven: Yale University Press, 1967), 118–19.

2. New York *Tribune*, 11 May 1899.

3. At the turn of the century, Americans used the word *environmentalist* to indicate a philosophy based on the idea that the environment influenced human development, what historians now call positive environmentalism. Thus the use of the term *environmentalist* to describe the antismoke reformers is anachronistic, since activists involved in the movement would not have used that word to describe themselves. However, these early environmental reformers defined pollution problems using the same terms and rhetoric that post–World War II environmentalists would later adopt. Antismoke reformers also used many of the same tactics as postwar environmentalists. In this sense, the environmentalist description is most accurate. For a discussion of positive environmentalism see Paul Boyer, *Urban Masses and Moral Order in America, 1820–1920* (Cambridge: Harvard University Press, 1978), 220–51.

4. See Gabriel Kolko, *The Triumph of Conservatism* (New York: Free Press, 1963), for an exaggerated argument concerning the conservative nature of progressive reform and reformers.

5. Robert Wiebe's *The Search for Order, 1877–1920* (New York: Hill & Wang, 1967) remains the best monograph covering the entire progressive era. Wiebe argues that a new middle class asserted its influence in the progressive era, exchanging new, urban values for obsolete traditional beliefs. Samuel Hays's *The Response to Industrialism, 1885–1914* (Chicago: University of Chicago Press, 1957) also offers important insight into the impulse to organize in industrializing America. Nell Irvin Painter, in *Standing at Armageddon* (New York: Norton, 1987), argues that in the late 1800s the middle

class feared revolution from below above all else (see p. xii). But the greatest fear of middle-class reformers was not socialism, the government control of the economy, but just the opposite, no regulation of the economy. Chaos, not control, remained the greatest concern for the urban middle class through the early 1900s.

6. Poster reprinted in James West Davidson et al., *Nation of Nations: A Narrative History of the American Republic, Vol. II: Since 1865* (New York: McGraw-Hill, 1994), 798.

7. Historian Samuel Hays has concluded that before World War II "environmental civic action was limited," apparently ignoring the long and vital crusades to improve water supplies, garbage removal, sewage removal, park systems, and city air. Efforts to pave streets, bury power lines, and plant street trees were also part of this major effort to improve the urban environment in the progressive era and even preceding decades. See Samuel Hays, *Beauty, Health, and Permanence* (New York: Cambridge University Press, 1987), 72.

Robert Gottlieb correctly identifies the urban industrial roots of the environmental movement in the soil of the progressive era. Unfortunately, his work provides little discussion of the most active reform movements. Gottlieb also offers an inaccurate assessment of the smoke abatement movement, missing entirely the pre-professional reform efforts led by middle-class women. See *Forcing the Spring* (Washington: Island Press, 1993), 47–59.

8. In large part this paradigm comes from the work of historian Samuel Hays, whose *Beauty, Health, and Permanence* remains the most comprehensive scholarly work concerning the postwar environmental movement, and whose *Conservation and the Gospel of Efficiency: The Progressive Conservation Movement, 1890–1920* (Cambridge: Harvard University Press, 1959) has set the tone for the discussion of conservation since its publication. This paradigm is so prevalent that often histories of environmental activism leave out progressive-era urban reform altogether, largely because it does not fit well with the conservation narrative dominated by Gifford Pinchot, Theodore Roosevelt, and John Muir. See for example Roderick Nash, *American Environmentalism: Readings in Conservation History* (New York: McGraw-Hill, 1990). Historians Martin Melosi and Joel Tarr have been pioneers in studying progressive-era environmental reform, and their work provides much of the basis for these claims. See Melosi, *Garbage in the Cities* (College Station: Texas A & M University Press, 1981) and *Pollution and Reform in American Cities* (Austin: University of Texas Press, 1980); Joel Tarr, *The Search for the Ultimate Sink: Urban Pollution in Historical Perspective* (Akron: University of Akron Press, 1996).

9. See Stephen Fox, *American Conservation Movement: John Muir and His Legacy* (Madison: University of Wisconsin Press, 1991), and John Reiger, *American Sportsmen and the Origins of Conservation* (Norman: University of Oklahoma Press, 1986).

one The Vital Essence of Our Civilization

1. New York *Tribune*, 11 May 1902; New York *Times*, 13, 17 May 1902. For a detailed discussion of the anthracite strike of 1902 see Robert J. Cornell, *The Anthracite Coal Strike of 1902* (New York: Russell & Russell, 1957).

2. James MacFarlane, *The Coal Regions of America: Their Topography, Geology, and Development* (New York: D. Appleton, 1873), xvii.

3. Charles Barnard, *Chautauquan* 7 (1887): 269. Coal also served as the source of dozens of consumer products derived through chemical manipulation. Aside from the coal gas manufactured in producer plants and consumed for lighting, heating, and cooking fuel, chemists used the many components in coal to derive benzine, naphtha, carbolic acid, dyes, and perfumes, among other marketable products ("The Magic of a Piece of Coal," *Cosmopolitan* 38 [1905]: 603–5). On the role of energy in industrial America see Sam H. Schurr and Bruce C. Netschert, *Energy in the American Economy, 1850–1975* (Baltimore: Johns Hopkins University Press, 1960); Martin Melosi, *Coping with Abundance: Energy and Environment in Industrial America* (Philadelphia: Temple University Press, 1985).

4. See the 20 January 1918 New York *Times* picture section for a moving photo of women and children searching through an ash heap in search of coal during World War I, when scarcity drove prices beyond the reach of many urban residents.

5. *Thirteenth Census of the United States* (1910) vol. 4, Population Occupation Statistics. In total the census lists 108,374 men as firemen tending stationary fires. In addition 76,381 firemen tended locomotive engines. See also *Historical Statistics of the United States Colonial Times to 1957* (1960), 358–59; John G. Clark, *Energy and the Federal Government* (Urbana: University of Illinois Press, 1987), 6–7. For a description of the effects of coal on mining families see David Alan Corbin, *Life, Work, and Rebellion in the Coal Fields* (Urbana: University of Illinois Press, 1981), particularly chapter 1, "Coal Is Our Existence."

6. Statistics from the Commerce Department reveal how important coal was as freight for some of the nation's most traveled lines. In 1914, for example, coal accounted for 51 percent of the Pennsylvania Railroad's total traffic and 56 percent of the Baltimore and Ohio's (W. P. Ellis, *Report of the Distribution Division*, Fuel Administration [1919], 7). Railroad companies owned considerable portions of the anthracite fields, but were much less interested in operating bituminous mines, where competition was rife and profitability was much less certain. For a discussion of the chaotic bituminous industry see William Graebner, "Great Expectations: The Search for Order in Bituminous Coal, 1890–1917," *Business History Review* 48 (1974): 49–72. For railroad involvement in anthracite mines see Cornell, *Anthracite Coal Strike of 1902*.

7. A third category of coal, lignites, formed only a small percentage of the coal market. Lignite, also known as brown coal, was the least valuable of coal types and garnered little demand.

8. For a discussion of coke see Joel A. Tarr, "Searching for a 'Sink' for an Industrial Waste: Iron-Making Fuels and the Environment," *Environmental History Review* 18 (1994): 12–13.

9. In 1901, the Geological Survey reported that standard Illinois lump coal averaged $2.80 per ton at the market, while large anthracite brought $6.65 per ton (United States Geological Survey, *Mineral Resources, 1901* [1902], 348). Melosi, *Coping with Abundance*, 30–31. Whereas clinkers were just an inconvenience for domestic users,

large consumers faced more serious problems from the residues of impure coal, as slag clogged grates and diminished drafts, thereby reducing boiler efficiency ("Smoke Prevention," *Journal of the Association of Engineering Societies* 11 [1892]: 294). See also Joel Tarr and Kenneth K. Koons, "Railroad Smoke Control: The Regulation of a Mobile Pollution Source" in *Energy and Transport*, ed. George H. Daniels and Mark Rose (Beverly Hills: Sage Publications, 1982), 71–92.

10. Clark, *Energy and the Federal Government*, 11. For a discussion of coal types available in St. Louis in the early 1890s, see "Smoke Prevention" (1892), 299–303.

11. United States Geological Survey, *Mineral Resources, 1910* (1911), 33. When Pennsylvania's oil and natural gas production are added to its coal tonnage, that state's leadership in energy production becomes even clearer. Behind Pennsylvania in coal production in 1910 were West Virginia (12.3 percent), Illinois (9.1 percent), Ohio (6.8 percent), and Indiana (3.7 percent).

12. Although some coal did travel considerable distances within the nation, the expense of shipping coal kept imports and exports of the fuel quite low. In 1899, for example, imports of coal equaled only .5 percent of the total production for that year. Exports for the same year equaled just 2.5 percent of production. These very low figures are representative for the period, at least until the late 1910s, when exports increased slightly. Thus, while measurements of consumption and production do differ, and the distinction between the two is important, given the limited American involvement in the international coal trade, for our purposes production can stand as a rough estimate of consumption, and vice versa (U.S.G.S., *Mineral Resources, 1899* [1900], 361–62).

13. U.S.G.S., *Mineral Resources, 1910* (1911), 78.

14. Ibid. In 1901, William Bryan estimated that in St. Louis those using Pocahontas coal would spend $.24 to evaporate 1000 pounds of water, while those using the much less expensive Mt. Olive coal would spend less than $.11 to do the same work (Bryan, "Smoke Abatement in St. Louis," *Journal of the Association of Engineering Societies* 27 [1901]: 221).

15. Sam H. Schurr and Bruce C. Netschert, *Energy in the American Economy, 1850–1975* (Baltimore: Johns Hopkins Press, 1960), 36, 85–108. See also Joseph A. Pratt, "The Ascent of Oil: The Transition from Coal to Oil in Early Twentieth-Century America," in *Energy Transitions: Long-Term Perspective*, ed. Lewis J. Perelman (Boulder: Westview Press, 1981).

16. Schurr and Netschert, *Energy in the American Economy*, 36, 125–39. See also John H. Herbert, *Clean, Cheap Heat: The Development of Residential Markets for Natural Gas in the United States* (New York: Praeger, 1992).

17. Schurr and Netschert, *Energy in the American Economy*, 36, 45–57.

18. Alfred D. Chandler, "Anthracite Coal and the Beginnings of the Industrial Revolution in the United States," *Business History Review* 46 (1972): 141–81; Martin Melosi, "Energy Transitions in the Nineteenth-Century Economy," in *Energy and Transport*, ed. Daniels and Rose; William F. M. Goss, "Smoke Responsibility Cannot Be Individual-

ized," *Steel and Iron* 49, pt. 1 (1915): 224. Coal not only aided in the growth of cities, it largely determined the location of the nation's most successful towns. Ore traveled all the way from northern Michigan and Minnesota, and men came from as far away as Russia and Poland to help burn coal in western Pennsylvania.

19. D. T. Randall, "The Government Fuel Investigation and the Smoke Problem," *Power and the Engineer* 29 (1908): 101; Melosi, *Coping with Abundance*, 33.

20. Schurr and Netschert, *Energy in the American Economy*, 508; "The Smoke Nuisance," *Medical Record* 69 (1906): 392; "Smoke, Dust and Gas," *The Sanitarian* 46 (1900): 188–89. Some coal consumers in eastern cities resisted the switch to bituminous coal for several reasons. First, anthracite was a more efficient fuel, and although it cost more per ton, depending on the price of the two fuels, it could actually do more work per dollar than its less expensive rival. Second, even when soft coal made economic sense, public pressure concerning new smoke nuisances created by the switch convinced some industries to stick with the more expensive anthracite. In 1897 Samuel Vauclain of the Baldwin Locomotive Works in Philadelphia expressed his company's dissatisfaction with a switch to soft coal. "It soon became evident that to continue the use of soft coal would not only call forth the protests of our immediate neighbors but also the opposition of the city fathers," Vauclain said. His company found a workable compromise when it began to burn a mixture of soft and hard coal which cut the smoke discharge and saved the company money in fuel expenses ("The Smoke Nuisance and Its Regulation," *Journal of the Franklin Institute* 143 [1897]: 419).

21. "The World's Coal Production" *Iron Age* 94 (1914): 1085; William Jasper Nicolls, *The Story of American Coals* (Philadelphia: Lippincott, 1897), 388. American mines first outproduced their British counterparts in 1899, when the United States produced nearly 254 million tons and Britain produced less than 247 million tons. American production had outpaced German production since 1871, making the United States the second largest producer in the world for the twenty-eight years before it surpassed Great Britain (U.S.G.S., *Mineral Resources, 1901* [1902], 310–13). See also Edward Atkinson, "Coal Is King," *Century Magazine* 55 (1898): 828–30.

22. Henry Obermeyer, *Stop That Smoke!* (New York: Harper & Brothers, 1933), 9–10.

23. Chester G. Gilbert, "Coal Products," *United States National Museum Bulletin* 102, pt. 1 (1917): 3; Chester G. Gilbert and Joseph E. Pogue, "Power: Its Significance and Needs," *National Museum Bulletin* 102 (1917): 7; Charles Barnard, "Rocks as Civilizers," *The Chautauquan* 7 (1887): 269; Gifford Pinchot, *The Fight for Conservation* (New York: Doubleday, Page, 1910), 43. Frank Ray, a professor of mining engineering at the Ohio State University, agreed that the importance of coal could not be overestimated: "It is the father of modern civilization," he wrote (Ray, "The Ohio Coal Supply and Its Exhaustion," Ohio State University Engineering Experiment Station *Bulletin* No. 12, 1914).

24. Robert W. Bruere, *The Coming of Coal* (New York: Association Press, 1922), 2, 10–12.

25. Birmingham *Age-Herald*, 1 February, 30 January 1913.

26. Chicago *Record-Herald*, 26 April 1909; Booth Tarkington, *Growth* (New York: Doubleday, Page, 1927), 319–20. *The Turmoil* originally appeared serialized in *Harper's* in late 1914 and early 1915 and became part of the larger *Growth* when it appeared with *The Magnificent Ambersons* and *National Avenue* in 1927.

27. William M. Barr, "The Smoke Nuisance," *Journal of the Association of Engineering Societies* 1 (1882): 401.

28. New York *Tribune*, 14 June 1902; New York *Sun*, 10 June 1902; New York *Times*, 4, 15 June 1902.

29. New York *Times*, 11 June 1902; New York *Tribune*, 14 June 1902. New York was not the only city so adversely affected by the strike. All eastern cities heavily dependent on anthracite saw a dramatic increases in fuel costs. Other cities also experienced heavy smoke. In Washington, D.C., authorities could not enforce a strong antismoke law due to a lack of anthracite coal, and by late September the Washington monument closed due to a shortage of coal to operate the elevator (New York *Times*, 22, 24 September 1902).

30. New York *Times*, 15 June 1902; New York *Tribune*, 5 June 1902.

31. New York *Times*, 22 May 1902; New York *Tribune*, 4 June 1902.

32. New York *Times*, 19 May, 12 December 1902.

33. New York *Tribune*, 28 May 1902; New York *Times*, 14, 27 June 1902.

34. New York *Tribune*, 10 June 1902. Under the headline "City's Thick Smoke Pall," this article announced that the Delaware, Lackawana, and Western Railroad had not sold anthracite since the strike began. The other sources are New York *Tribune*, 24 May 1902; New York *Times*, 25 July, 15 August, 27 September, 9 October 1902.

35. *The Nation* 75 (October 16, 1902): 298; New York *Times*, 16 October 1902. New York resumed the enforcement of the smoke ordinance on 15 November 1902, when the price of anthracite had receded to $6.50 per ton (New York *Times*, 15 November 1902).

36. The New York *Tribune* referred to the city's "black shroud" in a headline, 8 June 1902.

37. Peter Brimblecombe, "Attitudes and Responses towards Air Pollution in Medieval England," *Journal of the Air Pollution Control Association* 26 (October 1976): 941–45; New York *Tribune*, 14 June 1902. *Scientific American* also published a brief primer on how to fire bituminous coal efficiently, including information that had long been common knowledge in the soft coal cities of the Midwest (*Scientific American* 87 [1902]: 182).

38. New York *Times*, 30 September 1902.

39. Marcus M. Marks, president of the Manhattan Borough, quoted in New York *Times*, 13 November 1913. For a review of environmental problems in the progressive decades see Martin Melosi, ed., *Pollution and Reform in American Cities, 1870–1930* (Austin: University of Texas Press, 1980).

40. In the decades surrounding the turn of the century, the idea of civilization garnered considerable attention in public discourse, particularly in connection with en-

vironmental issues. Writer Martha Bensley Bruere referred to "our stage of semi-civilization" in an article concerning smoke. According to Bruere, smoke was a disease from which the nation suffered. Only a complete cure could elevate the United States to a higher civilization ("A Cure for Smoke-Sick Cities," *Collier's* 174 [November 1, 1924], 28). The political economist Simon Patten discussed at length the requirements for a higher civilization, focusing on the destruction of poverty, but also on the improvement of the human environment. In Patten's vision of a higher civilization, "all water will be as pure as that flowing from a spring, light and air will be as clear in the city as they are in the mountains, and the streets will be as clean and safe and honest as the home" (*The New Basis of Civilization* [New York: Macmillan, 1907], 196).

two Living with Smoke

1. For descriptions of nineteenth-century American urban environments see Martin Melosi, ed., *Pollution and Reform in American Cities, 1870–1930* (Austin: University of Texas Press, 1980), 3–23; and Melosi, *Garbage in the Cities* (College Station: Texas A & M Press, 1981), 16–20. For a general description of cities in last decades of the 1800s see Sam Bass Warner Jr., *The Urban Wilderness* (New York: Harper & Row, 1972), 85–112.

2. Waldo Frank, *Our America* (New York: Boni & Liveright, 1919), 117. For a lengthy discussion of outsiders' impressions of Chicago's smoke see William Cronon, *Nature's Metropolis* (New York: W. W. Norton, 1991), "Prologue: Cloud over Chicago," 5–19.

3. Letter to the editor by Frangipani Soot, Milwaukee *Sentinel*, 28 October 1888. The Scottish author William Archer, during a visit in 1900, compared Chicago's smoke to a thundercloud. See Bessie Pierce, ed., *As Others See Chicago* (Chicago: University of Chicago Press, 1933), 408–11.

4. New York *Tribune*, 24 April 1899, 13 December 1898, 27 April 1899.

5. "Pittsburgh's Smoke Abatement Exhibit," *Survey* 31 (1913): 1. St. Louis also conducted a soot fall study at the encouragement of the Mellon Institute. Directed by faculty at Washington University and funded by the Women's Organization for the Abatement of Smoke in St. Louis, the study collected soot from twelve spots in the city from April 1916 through March 1917. The study revealed that on average 812 tons of soot fell per square mile in St. Louis that year, for a total of 49,870 tons for the entire city (Ernest L. Ohle and Leroy McMaster, "Soot-Fall Studies in Saint Louis," *Washington University Studies* 5, pt. 1 [1917]: 7).

6. Soot fall statistics for Cincinnati appear in the Williams directory in a brief description of the city which preceded the listing of citizens, giving some indication of seriousness of the smoke problem in that city (*Williams' Cincinnati Directory* [1916], 12–13). Many of the nation's smokiest cities lay in river valleys, and their nearest suburbs rested on the surrounding hills. Pittsburgh, Cincinnati, St. Louis, Cleveland, and to some degree Milwaukee, all fit this description.

7. See Martin Melosi, *Coping with Abundance* (Philadelphia: Temple University Press, 1985), 32. Temperature inversions exacerbated the smoke problem in many cities, including Salt Lake City and Butte, where air pollution took on a more toxic quality due to the abundance of smelters, which emitted more than just coal smoke. But in most industrial cities, smoke was a nuisance in almost any kind of weather. London became internationally famous for its inversions, known as "fogs," which occasionally became so dense as to be deadly. In the early 1900s, these deadly smoke fogs took on a new name, "smogs." London's most deadly smogs occurred in 1873, 1880, 1892, and 1952. See Eric Ashby and Mary Anderson, *The Politics of Clean Air* (New York: Oxford University Press, 1981), and Peter Brimblecombe, *The Big Smoke* (New York: Methuen, 1987). Quotation from James Parton, "Pittsburgh," *Atlantic Monthly* 21 (1868): 21.

8. The very definition of smoke included the word *visible*, and for the most part city residents expressed concern for only the visible portions of smokestack and chimney emissions. City ordinances relied on the terms *dense* or *black* to describe prohibited emissions, and commentators most often remarked about the opaque nature of air pollution and of soot, the visible portion of emissions which settled out of the air. This does not suggest, however, that urbanites did not understand the complex composition of emissions. Antismoke activists also remarked on the negative effects of sulphur in coal smoke, particularly on plants and building surfaces. And, smoke itself was not the only air pollution in turn-of-the-century industrial cities, where emissions and vapors of all kinds faced little regulation. In 1907, Amos Price of Youngstown, Ohio offered commentary about a trip to Cleveland which suggests the seriousness of air pollution problems: "I was in Cleveland in the locality of Broadway and the Erie railroad, where enough crude oil arose from the tanks to mingle with the smoke for a while, then the compound fell in chunks big enough to be felt, the only means of recognition through the murky atmosphere, except by sight at very close range" (*Power* 27 [1907], 120).

Residents near smelters well understood that emissions from those stacks contained more than just unburned carbon. Unlike most coal fires, smelter fires emitted highly toxic gases and arsenic particulates. In cities with large smelters, such as Butte, Montana, the effects of air pollution were devastating. Still, in most of the nation's cities the effort to clean urban air focused on visible emissions, i.e., unburned carbon. For a study of the effort to control smelter smoke see Donald MacMillan, "A History of the Struggle to Abate Air Pollution from Copper Smelters of the Far West, 1885–1933" (Ph.D. diss., University of Montana, 1973).

9. Cincinnati Board of Health, *Annual Report* (1886), 55, 19. See David Stradling, "The Price of Blue Skies: The Anti-Smoke Movement in Cincinnati, 1868–1916" (Master's thesis, University of Wisconsin–Madison, 1991). For the state of the understanding of the health effects of smoke see Oskar Klotz and William Charles White, eds., "Papers on the Influence of Smoke on Health," *Mellon Institute Smoke Investigation Bulletin No. 9* (1914). The Mellon research made clear the influence of smoke on

pneumonia and other respiratory diseases. Throughout this era Americans relied heavily on the research of British and German epidemiologists for data concerning the health effects of smoke and soot.

10. Chicago *Record-Herald*, 18 April 1911, 21 December 1905; Cleveland Chamber of Commerce, "Report of the Municipal Committee on the Smoke Nuisance" (1907); William H. Wilson, *The City Beautiful Movement* (Baltimore: Johns Hopkins University Press, 1989), passim. For a discussion of the role of smoke in City Beautiful conceptions of city planning see John O'Connor, "The City Beautiful and Smoke," *Pittsburgh Bulletin* 65, no. 4 (1912): 9. City Beautiful plans generally revolved around one central idea: city vistas should be impressive, both grand and beautiful. Obviously smoke could quite easily spoil these plans.

11. In 1904, the Chicago *Record-Herald* lamented the rapid soiling of the city's new federal building. "May the day soon come when it will be possible to scrub the new federal building with soap and water, with some prospect that the marble will remain white long enough to reward the public for the labor and expense of such a colossal cleansing" (*Record-Herald*, 21 September 1904). See John O'Connor, "The City Beautiful and Smoke," *Pittsburgh Bulletin* 65 (1912): 9, for a discussion of the civic center and White City ideals. For a detailed description of the White City see David F. Burg, *Chicago's White City of 1893* (Lexington: University of Kentucky Press, 1976).

12. J. E. Wallace Wallin, "Psychological Aspects of the Problem of Atmospheric Pollution," *Mellon Institute Smoke Investigation Bulletin No. 3* (1913): 36; "Art in War on Smoke," Chicago *Record-Herald*, 17 July 1908. Health officials gave considerable attention to the issue of sunlight, often attributing many of the problems of tenement districts to restricted light in apartments. Basement apartments garnered special attention as harborers of filth, disease, and depravity. One sanitary engineer, Charles F. Wingate, went so far as to claim that "deficiency of light is usually characterized by ugliness, rickets, and deformity, and is a fruitful source of scrofula and consumption in any climate" (*Sanitary News* 14 [1889]: 55). For an exposé on conditions in tenement basement apartments see Benjamin Flower, *Civilization's Inferno* (Boston: Arena, 1893).

13. Herbert H. Kimball, "The Meteorological Aspect of the Smoke Problem," *Mellon Institute Smoke Investigation Bulletin No. 5* (1913): 48. An abridged version of Kimball's work also appeared in the government publication *Monthly Weather Review* 42 (1914): 29–35.

14. Svante Arrhenius, "The Influence of the Carbonic Acid in the Air upon the Temperature of the Ground," *Philosophical Magazine* 41 (1896): 237–76; Charles Richard Van Hise, *The Conservation of Natural Resources* (New York: Macmillan, 1924; orig. pub. 1910), 33; "The Increasing Temperature of the World," *Scientific American* 107 (1912): 99. A decade before Arrhenius announced his findings, a German chemist, Clemens Winkler, wrote an essay entitled "The Influence of the Combustion of Coal upon Our Atmosphere," in which Winkler asked, "Is it possible that the carbonic acid which is produced in so great quantities by modern industry, may *not* be consumed by plants, but amassed gradually in our terrestrial atmosphere?" Although Winkler

admitted that he did not know the answer, he expressed faith in humanity's lack of power to influence such cosmic forces (*Open Court* 1 [1887]: 197–99).

15. Civic League of St. Louis, "The Smoke Nuisance" (1906), 5–6; "They Live in the Shadow of a Miniature, but Active Volcano," *Cleveland Press*, 6 January 1905. Smoke's effect on vegetation garnered enough concern to warrant the attention of an entire volume in the Mellon Institute's seminal investigation of Pittsburgh's smoke problem; see J. F. Clevenger, "The Effect of Soot in Smoke on Vegetation," *Mellon Institute Smoke Investigation Bulletin No. 7* (1913).

16. Paul Boyer, *Urban Masses and Moral Order in America* (Cambridge: Harvard University Press, 1978), 220–51. Boyer is most concerned, as were progressive reformers, with the effects of the dirty urban environment on the morality of the lower classes. Stanley Schultz also discusses "moral environmentalism" as it developed in the 1800s (*Constructing Urban Culture* [Philadelphia: Temple University Press, 1989], 112–14). Arthur Dudley Vinton's "Morality and Environment" offers a succinct summary of positive environmentalism. In describing a child raised in a New York City tenement district, Vinton concludes, "He has no choice but to become wicked," because of his environment (*Arena* 3 [1891], 567–77). Remaining quotation are from Wallin, "Psychological Aspects of the Problem of Atmospheric Pollution," 32; Charles A. L. Reed, "An Address on the Smoke Problem" (1905), 3; "Smoke Causes Crime," Cincinnati *Times-Star*, 23 September 1907. Nelson relied on the work of Dr. W. T. Talbot of New Hampshire for his conclusions about the relationship between smoke and crime.

17. "Women Complain of Smoke Evil," Milwaukee *Sentinel*, 10 November 1903. Many nineteenth-century authors discussed in detail the relationships among beauty, health, cleanliness, and morality. See for example Charles Loring Brace, *The Dangerous Classes of New York and Twenty Years' Work among Them* (New York: Wynkoop & Hallenbeck, 1872), esp. 56–69.

18. "Millions Spent in Cleaning Dirt from Buildings," Chicago *Tribune*, 20 May 1913. The *Tribune* expressed particular concern for the condition of the new buildings: "To see how quickly the shinning white wall is covered with a gray and slimy coating is to make it uncomfortable to imagine the condition of one's bronchial tubes and lungs, which do not happen to be lined with glazed terra cotta." The chief fellow of the Mellon Institute's study of the smoke problem, Raymond Benner, published an issue dedicated to this problem: "Papers on the Effect of Smoke on Building Materials," *Mellon Institute Smoke Investigation Bulletin No. 6* (1913).

19. *National Municipal Review* 6 (1917): 591–98. James Parton described at length Pittsburgh's smoke in a 1868 essay on that city. "All dainty and showy apparel is forbidden by the state of the atmosphere," he wrote, "and equally so is delicate upholstery within doors" (*Atlantic Monthly* 21 [1868]: 19).

20. "The Smoke Nuisance: Report of the Smoke Abatement Committee," Civic League of St. Louis (1906), 4–5.

21. O'Connor used reports from Pittsburgh businesses to estimate that 30 percent

of the artificial light used in that city was due to the unnatural darkness caused by smoke. O'Connor also noted that not only did smoke increase the demand for artificial light, but it also diminished its effectiveness by soiling the globes and shades that protected the lamps (John J. O'Connor Jr., "The Economic Cost of the Smoke Nuisance to Pittsburgh," *Mellon Institute Smoke Investigation Bulletin No. 4* [1913], 31).

22. Ibid. For the role of urban environmental problems in suburban development see Mark Rose, "'There Is Less Smoke in the District,'" *Journal of the West* 25, no. 2 (1986): 44–54. By the late 1910s there was one exception to this general statement about land values near railroads: the smoke in midtown Manhattan had been greatly reduced due to the electrification of Grand Central Station and the elevated railroads that served the area. See Kurt C. Schlichting, "Grand Central Terminal and the City Beautiful in New York," *Journal of Urban History* 22 (1996): 332–49.

23. O'Connor, "The Economic Cost," 40–41.

24. "Smoke Cost Fabulous," Chicago *Record-Herald*, 18 November 1909. Wilson disclosed his estimates at a meeting of the American Civic Association in Cincinnati.

25. In 1909, the year of the Cleveland estimate, the average nonfarm wage earner made $594, assuming full-time employment. Unskilled laborers averaged less than $11 in weekly earnings. Of course the single per-family cost derived by the Chamber of Commerce did not reflect the real variance in damage. Since working-class families tended to live in smokier areas of cities, they undoubtedly incurred greater costs than did wealthier families who lived in relatively clean neighborhoods. Paul H. Douglas, *Real Wages in the United States, 1890–1926* (New York: Houghton Mifflin, 1930), 177, 391.

26. Although this estimate represented less than 2 percent of the nation's GNP in 1909, the cost was not evenly distributed around the nation and was born primarily by those who lived in cities that consumed significant quantities of soft coal. Quotations are from Cleveland Chamber of Commerce, "Report of the Committee on Smoke Prevention" (1909), 4–10; *Industrial World* 45, pt. 2 (1911): 1321; O'Connor, "The Economic Cost," 8–10; Chicago *Record-Herald*, 18 November 1909. These estimates of the cost of smoke to cities were widely published. See, for example, *Outlook* 93 (1909): 6. *The Iron Age, Power, Industrial World*, and other trade magazines paid particularly close attention to the economic and engineering aspects of the smoke problem. For wage statistics see United States Bureau of the Census, *Historical Statistics of the United States Colonial Times to 1957* (1960).

27. These estimates may be called conservative due to their failure to add costs of smoke on health, which were real and considerable. None of the estimates discussed the costs of lost work time, shortened working lives, hospital or doctors' fees. In his attempt to determine the cost of smoke to Pittsburgh, the Mellon Institute's C. W. A. Veditz corresponded with Bird in Chicago and Edward Jerome in Cincinnati. Jerome noted that no research had been conducted to determine the actual cost of smoke in his city. Veditz replied, "All the figures which have thus far been published, with regard to the financial losses attributable to smoke are simply guesses, often of the

vaguest and most indefinite kind." Veditz worried that the lack of scientific research into the costs of smoke would limit the influence of such figures with the public. See Jerome to Veditz, 19 March 1912; Veditz to Jerome, 22 March 1912, Smoke Investigation Activities MSS, series 1, folder 21, University of Pittsburgh Library.

28. *Sanitary News* 18 (1891): 202.

29. New York *Tribune*, 11 December 1898, 8 December 1893; Chicago *Record-Herald*, 13 April 1907.

30. Wallin, "Psychological Aspects of the Problem of Pollution," 8–9. Although Wallin was adamant in his assertions concerning smoke and mental health, his was not the most extreme view. In the same year J. B. Stoner, of the United States Public Health Service, wrote, "Women living in sunless gloomy homes and attired in somber clothes are also prone to be irritable, to scold and whip their children and to greet their husbands with caustic speech" (*Military Surgeon* 32 [1913]: 373). As early as 1882 a London physician had linked smoke to "general depression," a linkage that remained assumed and casually mentioned throughout the progressive decades (*Sanitary News* 1 [15 December 1882]: 38).

31. Upton Sinclair, *The Jungle* (New York: Signet New American Library, 1960; orig. pub. 1906), 29–30, 34.

32. Ironically, the Marshall Field Company had itself come under attack for violating the smoke ordinance in 1891. Field was not the only smoke producer to express deep concern for smoke pollution. In a nation so dependent on coal, few urban residents could claim innocence in the production of smoke. See *Marshall Field & Company v. City of Chicago* 44 Ill App 410 (1892).

33. "The Smoke testimony of One Week," Chicago *Record-Herald*, 25 February 1907. Breckenridge published his study later that year in an oft-cited report: "How to Burn Illinois Coal without Smoke," *University of Illinois Engineering Experiment Station, Bulletin No. 15* (August 1907).

34. "The Atmospheric Crisis with which Civilization Is Threatened," *Current Literature* 43 (1907): 331. See Melosi, ed., *Pollution and Reform in American Cities*, for successes in garbage and sewage removal and water supply.

35. See Nelson Blake, *Water for the Cities* (Syracuse: Syracuse University Press, 1956), 144–71, for a narration of the building of the Croton Aqueduct. See Louis P. Cain, *Sanitation Strategy for a Lakefront Metropolis* (Dekalb: Northern Illinois University Press, 1978), for a description of the extensive efforts to keep Chicago's water supply pure.

36. See for example, Robert C. Benner, "Why Smoke Is an Industrial Nuisance," *Iron Age* 91 (1913): 135–38. Although industry also created some serious water pollution problems, in the decades before World War I urban residents expressed much greater concern for domestic water pollution. Health officials had long associated sewage with disease, but the health effects of industrial wastes were less well understood. Diluting industrial pollutants in large bodies of water, i.e., dumping untreated waste into waterways, remained the dominant disposal practice through World War

I. See Craig Colton, "Creating a Toxic Landscape," *Environmental History Review* 18 (1994): 86–88; and Joel Tarr, "Industrial Wastes, Water Pollution, and Public Health, 1876–1962," in *The Search for the Ultimate Sink* (Akron: University of Akron Press, 1996), 354–84.

When industrial water pollution did spark public demands for relief, powerful industries escaped effective control from incompetent or inactive local governments. Thus those water pollution problems associated with industry received little effective regulation before the 1920s. Meanwhile, the urban problem of sewage disposal garnered considerable municipal attention in the struggle for improved public health. See Andrew Hurley, "Creating Ecological Wastelands," *Journal of Urban History* 20 (1994): 349–60. Although industrial water pollution problems only increased during World War I and through the next decades, as both the chemical and petroleum industries grew, concern for waterborne human wastes still exceeded those for industrial wastes. Andrew Hurley relates the story of Chicago's effort to control the pollution of Lake Michigan from the massive Gary Works in 1943, which culminated in an agreement requiring the plant to treat its human waste but not its industrial wastes, which it continued to dump into the lake and the Grand Calumet River. See Hurley, *Environmental Inequalities* (Chapel Hill: University of North Carolina Press, 1995), 42–43.

37. For a discussion of the cholera epidemics of 1832, 1849, and 1866, see Charles Rosenberg, *The Cholera Years* (Chicago: University of Chicago Press, 1962). Cities faced another type of crisis as they attempted to solve their water supply problems: fire. In fact a 1835 fire in New York City finally forced action of the Croton Aqueduct (Blake, *Water for the Cities*, 143).

38. This is not true of London, where occasional intense smoke fogs (smogs) led to the deaths of thousands of Londoners in the 1800s. See Ashby and Anderson, *The Politics of Clean Air*. Donora, Pennsylvania also experience a deadly smoke fog in 1948. See Lynne Page Snyder, "'The Death-Dealing Smog over Donora, Pennsylvania,'" *Environmental History Review* 18 (1994): 117–39.

39. "Smoke a Sanitary, Not Only an Esthetic Nuisance," *American Medicine* 3 (1902): 800.

three The Movement Begins

1. Letter to the Editor, Pittsburgh *Gazette*, 7 November 1823; "Pittsburgh," *Atlantic Monthly* 21 (1868): 18. Willard Glazier quoted in Charles Glaab, *The American City: A Documentary History* (Homewood: Dorsey Press, 1963), 236. For other early accounts of Pittsburgh's smoke see John Duffy, "Smoke, Smog, and Health in Early Pittsburgh," *Western Pennsylvania Historical Magazine* 45 (1962): 93–106.

2. Cleveland *Daily Leader*, 7 May 1869, 26 October 1871.

3. For limited nature of antismoke efforts before the 1890s see Lucius Cannon, *Smoke Abatement* (St. Louis: St. Louis Public Library, 1924), 210–11, 231–32; John

O'Connor, "The History of the Smoke Nuisance and of Smoke Abatement in Pittsburgh," *Industrial World* 47, pt. 1 (1913): 353–54.

4. In 1869, Pittsburgh also specifically prohibited the construction of coke ovens within the city limits. The city appears not to have enforced that provision either, as more than a hundred coke ovens continued to operate within the city through the late 1800s (Joel A. Tarr, "Searching for a 'Sink' for an Industrial Waste: Iron-Making Fuels and the Environment," *Environmental History Review* 18 [1994]: 15).

5. Augustus A. Straub, "Some Engineering Phases of Pittsburgh's Smoke Problem," *Mellon Institute Smoke Investigation Bulletin No. 8* (1914), 12. Cincinnati Smoke Inspector Edward Jerome recited this version of the failure of the first antismoke laws in Cincinnati in 1915. See "Proceedings of the Tenth Annual Convention of the International Association for the Prevention of Smoke" (Cincinnati, 1915), 36. For a discussion of the ordinance see City of Cincinnati, *Annual Reports of the Departments, 1881*, 668–69. For judicial support of Chicago's 1881 ordinance see *Harmon v. City of Chicago*, 110 Ill. 400 (1884); *Marshall Field & Company v. City of Chicago*, 44 Ill. App. 410 (1892). For a description of the earliest efforts to abate smoke in Chicago see Christine Meisner Rosen, "Businessmen against Pollution in Late-Nineteenth-Century Chicago," *Business History Review* 69 (1995): 351–97.

6. Milwaukee *Daily Sentinel*, 11 October 1879; 6, 13, 16 November and 29 December 1880; 26 January 1881.

7. *Huckenstine's Appeal*, 70 PA 107 (1871). For a summary of smoke nuisance law see H. G. Wood, *A Practical Treatise on the Law of Nuisances in Their Various Forms* (Albany: John D. Parsons, 1875), 470–514. In another-oft cited Pennsylvania case, *Galbraith v. Oliver* 3 Pitts. 79 (1867), the court asked the difficult question, "How much atmosphere has a man a right to have preserved in its purity for his use?" This court answered only so much as "is necessary for his personal health and comfort, and the safety of his property." Later courts would give a more relative answer, differentiating urban and rural dwellers. For additional discussions of nineteenth-century nuisance case law see Jan G. Laitos, "Continuities from the Past Affecting Resource Use and Conservation Patterns," *Oklahoma Law Review* 28 (Winter 1975): 60–96; and Harold W. Kennedy and Andrew O. Porter, "Air Pollution: Its Control and Abatement," *Vanderbilt Law Review* 8 (1955): 854–77.

8. *Campbell et al. v. Seaman*, 63 NY 577 (1876). In this case the judge found in favor of the plaintiff, the owner of a large suburban estate south of Albany who demanded an injunction against a highly polluting brick manufacturer near his property. *Louisville Coffin Company v. Warren, &c.*, 78 KY 403–4 (1880); *Rhodes v. Dunbar*, 57 PA 286 (1868). See also *Richard's Appeal*, 57 PA 112 (1868), in which the Pennsylvania Supreme Court found in favor of the Phoenix Iron Company even after noting that the firm's emissions did cause harm to a neighbor's property, arguing that good iron was "an article of such prime necessity" that some harm could be justified.

9. *Appeal of the Pennsylvania Lead Company*, 96 PA 116, 127 (1880); *People v. The Detroit White Lead Works*, 82 Mich 471 (1890). Complainants against smelter smoke were

not guaranteed relief by court injunction simply due to the unhealthfulness of the emissions, however. Smelters in rural areas enjoyed considerable legal protection. In a case against the Tennessee Copper Company in 1904, for example, a Tennessee court acknowledged the damage done to area farms, particularly on crops and timber, and even to the wives of farmers, who complained of coughs and headaches. But while acknowledging the damage, the court did not issue an injunction against the polluter, noting that the smelter could not remove itself to a more remote place. The opinion asked, "Shall the complainants be granted, in the way of damages, the full measure of relief to which their injuries entitle them, or shall we go further, and grant their request to blot out two great mining and manufacturing enterprises, destroy half the taxable values of a county, and drive more than 10,000 people from their homes?" Clearly courts could deprive rural residents of their right to clean air if they should live near valuable industrial enterprises (*Madison v. Copper Company*, 113 Tenn 366 [1904]).

10. *Robinson v. Baugh*, 31 Mich. 290 (1875). See also *Ross v. Butler*, 19 NJ Eq. 294 (1868). Not all attempts to keep industry out of residential areas succeeded. See, for example, *Adams v. Michael*, 38 Md. 123 (1873), in which a Maryland court refused to prevent the construction of a factory "in the vicinity of certain valuable dwelling houses."

11. Although businessmen's clubs studied the smoke problem, often forming special committees for the purpose, for the most part businessmen's organizations, particularly the Chamber of Commerce, became active in the fight against smoke only after the turn of the century. Engineering clubs also studied the smoke problem, but rarely with the intention of influencing public policy in the 1890s.

12. John O'Connor, "The History of the Smoke Nuisance and of Smoke Abatement in Pittsburgh," *Industrial World* 47, pt. 1 (1913): 353–54; *Pittsburgh: Its Commerce and Industries, and the Natural Gas Interest* (Pittsburgh: George B. Hill, 1887), 41; *Sanitary News* 18 (1891): 53, 388; *Proceedings of the Engineers' Society of Western Pennsylvania* 8 (1892): 22–49, 51–73, 299–323. Representatives of the women's organization attended the society's meetings in February and March, and at both gatherings brought male representatives to argue their side. Several members, including the president, Alfred Hunt, offered support for the women's arguments against smoke, while others maintained that smoke was largely necessary and probably good for the city. The Women's Health Protective League was also identified as the Ladies' Health Protective Association.

13. Suellen M. Hoy, "'Municipal Housekeeping': The Role of Women in Improving Urban Sanitation Practices, 1880–1917," in *Pollution and Reform in American Cities*, ed. Martin Melosi (Austin: University of Texas Press, 1980), 173–98; Hoy, *Chasing Dirt: The American Pursuit of Cleanliness* (New York: Oxford University Press, 1995), 72–86; Angela Gugliotta, "Women and Anti-Smoke Activism in Pittsburgh, 1880–1920" (paper presented before the biennial meeting of the American Society for Environmental History, Baltimore, Md., March 1997). For a discussion of the influence of gender in reform politics see Maureen A. Flanagan, "Gender and Urban Political Reform," *American Historical Review* 95 (1990): 1032–50. For a discussion of the development of

women's clubs and municipal housekeeping see also Karen J. Blair, *The Clubwoman as Feminist* (New York: Holmes & Meier, 1980); Marlene Stein Wortman, "Domesticating the Nineteenth-Century American City," *Prospects* 3 (1977): 531–72; Mrs. T. J. Bowlker, "Woman's Home-Making Function Applied to the Municipality," *American City* 6 (1912): 863–869; and Mary Ritter Beard, *Woman's Work in Municipalities* (New York: D. Appleton, 1915), esp. 45–96. Women in other cities also formed Ladies' Health Protective Associations. See for example New York City, where women organized in 1884 and pledged to take action to ensure enforcement of sanitary laws and to compel the passage of amendments to those laws that were insufficient ("Charter and By-Laws of the Ladies' Health Protective Association of the City of New York," 1885).

14. *Proceedings of the Engineers' Society of Western Pennsylvania* 8 (1892), 31, 44–46. Rend went so far in his defense of smoke as to claim that carbon deposits in the lungs actually purified air as it passed into the blood. Rend had made similar statements for public consumption in Chicago; while speaking before the Union League Club early in 1892, he exclaimed, "Smoke is the incense burning on the altars of industry. It is beautiful to me. It shows that men are changing the merely potential forces of nature into articles of comfort for humanity. . . . Smoke means manufactures and manufactures have built our city" (as quoted in Rosen, "Businessmen against Pollution in Late-Nineteenth-Century Chicago," 385–86).

15. O'Connor, "The History of the Smoke Nuisance," 353–54; Robert Dale Grinder, "From Insurgency to Efficiency: The Smoke Abatement Campaign in Pittsburgh before World War I," *Western Pennsylvania Historical Magazine* 61 (1978): 187–202.

16. "Smoke Prevention: Report of the Special Committee on Prevention of Smoke, Presented to Engineers' Club of St. Louis," *Journal of the Association of Engineering Societies* 11 (1892): 322–23; "To Stop the Smoke," St. Louis *Post-Dispatch*, 22 January 1893; Cannon, *Smoke Abatement*, 212; William H. Bryan, "Smoke Abatement in St. Louis," *Journal of the Association of Engineering Societies* 27 (December 1901): 215–31.

17. Charles F. Olney, "My Dear Friends . . . ," undated speech from 1897 or 1898, Charles Fayette Olney Papers, 1894–1898, Oberlin College Archives. Cleveland's 1882 ordinance declared smoke a nuisance and created the position of smoke inspector. For two years the city made some progress, as the inspector aided businesses in equipping boilers with mechanical stokers and steam jets, but in 1884 the city repealed the ordinance (*Industrial World* 41, pt. 2 [1907]: 860–61, indicated as X and XI in the volume). Charles Olney became an active member in Cleveland's City Beautiful movement through his support of a city plan that would group public buildings in a grand civic center. See Elroy McKendree Avery, *A History of Cleveland and Its Environs*, vol. 1 (Chicago: Lewis Publishing, 1918), 467. For a review of the legal aspect of the Cleveland abatement effort see Cannon, *Smoke Abatement*, 245–47.

18. Chicago residents also created a Society for the Prevention of Smoke in 1891, but with a narrow goal. Several civic-minded business elites with financial interests in the 1893 World's Fair hoped to impress the world with the White City and Chica-

go as a whole. Abating Chicago's smoke became an important part of their improvement efforts since members feared that soot would prevent the White City from becoming the model for the clean, planned, healthful city of the future. The Society lasted only until 1893 (Rosen, "Businessmen against Pollution in Late-Nineteenth-Century Chicago," 351–87). See also Chicago Association of Commerce, *Smoke Abatement and Electrification of Railway Terminals in Chicago* (1915), 82–96.

19. "Women Complain of Smoke Evil," Milwaukee *Sentinel*, 10 November 1903.

20. "Smoke Nuisance and Health," St. Louis *Post Dispatch*, 8 May 1899; "Smoke Causes Crime," Cincinnati *Times-Star*, 23 September 1907; *Commercial Tribune*, 25 December 1908; Cincinnati *Post*, 14 December 1909; "Smoke, Fog, and Health," Chicago *Record-Herald*, 7 April 1904; Samuel W. Skinner, "Smoke Abatement" (paper read before the Optimist Club, Cincinnati, 1899, 19).

21. As quoted in R. Dale Grinder, "The Battle for Clean Air: The Smoke Problem in Post–Civil War America," in *Pollution and Reform in American Cities*, ed. Melosi, 86.

22. See, for example, Bernard J. Newman, "The Home of the Street Urchin," *National Municipal Review* 4 (1915): 587–93.

23. Engineers' Club of St. Louis, *Journal of the Association of Engineering Societies* 11 (1892): 322. Engineering journals also indicated health concerns as the primary incentive for smoke prevention research and development, often briefly and vaguely citing health issues before delving into more technical matters. In one such article in the *Sibley Journal of Engineering*, the author, writing well outside his area of expertise, concluded that "the danger of breathing the carbon-laden atmosphere is easily understood." The author arrived at this conclusion even though medical science had not yet determined how or even if smoke significantly affected health ("Smoke Prevention in Boiler Furnaces," *Sibley Journal of Engineering* 13 [1899]: 310).

24. "Pure Air and Clear Skies," *Outlook* 61 (1899): 106–7; "A Note of Warning," *Outlook* 60 (1898): 898. See also "The Smoke Nuisance," *Outlook* 60 (1898): 1000.

25. Oskar Klotz and William Charles White, eds., "Papers on the Influence of Smoke on Health," *Mellon Institute Smoke Investigation Bulletin* 9 (1914): 8; *Journal of the Franklin Institute* 143 (1897): 402, and 144 (1897): 37. See Charles Rosenberg, *The Cholera Years* (Chicago: University of Chicago Press, 1962), for descriptions of the use of smoke as a disinfectant.

26. Arguments concerning the disinfectant quality of smoke were particularly popular in smelter cities, such as Butte, Montana, where physicians could claim that highly sulphurous emissions actually improved public health. Sulphur was a known disinfectant, and physicians argued that smelter smoke killed germs throughout the community, reducing the incidence of certain types of diseases. The defense of smoke as disinfectant remained a potent argument in smelter cities into the twentieth century. See Duane Smith, *Mining America* (Lawrence: University of Kansas Press, 1987), 76–78; Donald MacMillan, "A History of the Struggle to Abate Air Pollution" (Ph.D. diss., University of Montana, 1973), 79–80.

27. *Franklin Institute* 144 (1897): 23. The president of Philadelphia's Board of

Health, William Ford, hedged about the lack of scientific proof of smoke's deleterious effects during the Franklin Institute's study. While listing some of his concerns about smoke Ford said, "from the evidence collected there can hardly be a reasonable doubt that there is from this cause a possibility of injury to health" (*Franklin Institute* 144 [1897]: 31).

28. John H. Griscom, *The Uses and Abuses of Air: Showing Its Influence in Sustaining Life, and Producing Disease* (New York: J. S. Redfield, 1848), passim.

29. George Derby, *An Inquiry into the Influence of Anthracite Fires upon Health* (Boston: A. Williams, 1868), esp. 52, 62–66. For a discussion of waste disposal "sinks" see Tarr, "Searching for a 'Sink,'" 9–10.

30. David Schuyler, *The New Urban Landscape* (Baltimore: Johns Hopkins University Press, 1986), 59–60; Sydney Dunham, "The Air We Breathe," *Chautauquan* 23 (1896): 145. Humanitarians concerned with the fate of poor urban children established Fresh Air Funds in many large cities as a means of alleviating at least some of the perceived negative effects of the tenement atmosphere. These funds, some dating back to the 1880s, allowed selected children to visit rural families for a number of days during the summer. Those who supported the funds thought "fresh air," even a few days' worth, might aid in the physical and moral development of these poor children. New York's *Tribune* sponsored the fund in that city (Harry W. Baehr Jr., *The New York Tribune since the Civil War* (New York: Dodd, Mead, 1936), 236. See also Willard Parsons, "The Story of the Fresh-Air Fund," *Scribner's Magazine* 9 (1891): 515–24.

31. Barr's tale gained some attention during the movement to improve air quality in the late 1960s, when it was reprinted. See James P. Lodge, ed., *The Smoke of London: Two Prophecies* (Elmsford: Maxwell Reprint Company, 1969). Barr was hardly the most popular of European fiction writers to use smoke in his/her work. Antismoke activists often referred to Charles Dickens's Coketown, as described in *Hard Times*. In his lasting criticism of the British industrial order and the stifling pragmatism of the new Brit, Dickens uses smoke throughout, often in majestic, fanciful terms, but more often to set the gloomy scene of Coketown: "Seen from a distance in such weather, Coketown lay shrouded in a haze of its own, which appeared impervious to the sun's rays. You only knew the town was there, because you knew there could have been no such sulky blotch upon the prospect without a town. A blur of soot and smoke, now confusedly tending this way, now that way . . . Coketown in the distance was suggestive of itself, though not a brick of it could be seen" (Dickens, *Hard Times* [New York: Oxford University Press, 1992; orig. pub. 1854], 146).

32. The authority to determine the health effects of smoke was only one of many areas in which women gradually lost their voice to scientifically trained physicians. By the turn of the century physicians had assumed considerable public stature, through professionalization, the emphasis on science, and a long battle against quackery. While women remained linked to health issues, particularly as they related to children and the home, trained physicians dominated public health issues. See Judith Leavitt's *Brought to Bed* (New York: Oxford University Press, 1986), which explores the

history of childbirth in America and the assertion of male physician's scientific knowledge over the traditional knowledge of females.

33. "The Sanitary Bearings of Smoke Nuisance," *Medical News* 64 (1894): 51.

34. "Smoke," *Journal of the American Medical Association* 44, pt. 2 (1905): 1619.

35. Ascher's findings found a broad audience in the United States and Europe. Even the *Engineering News* published a summary of Ascher's work (58 [1907]: 434–35). See also *Journal of the Royal Sanitary Institute* 28 (1907): 88–93. Ascher was not the only European scientist whose smoke research gained attention in the United States. See, for example, J. B. Cohen, "The Air of Towns," *Smithsonian Institution Annual Report, 1895*, 349–87; *Journal of the American Medical Association* 47 (1906): 41.

36. *Journal of the American Medical Association* 47 (1906): 385–86, and 49 (1907): 813–15. Schaefer's results appeared in *Boston Medical and Surgical Journal* 157 (1907): 106–10.

37. John Wainwright, "Bituminous Coal Smoke," *Medical Record* 74 (1908): 792. Medical findings concerning the health effects of smoke received considerable attention in the popular press. See for example the Cincinnati *Times-Star*, 11 November 1908, which reviewed Wainwright's findings. "The Value of Sunshine," *Sanitary News* 14 (1889): 70–71; Hollis Godfrey, "The Air of the City," *Atlantic Monthly* 102 (1908): 65.

38. Historian Christine Rosen has identified the depression of the mid-1890s as an important factor in the demise of the early antismoke campaign in Chicago, but she also explains the role of several other factors. Rosen argues that the Chicago Society for the Prevention of Smoke gained limited support for its campaign because the public remained skeptical of the elite businessmen's organization. Many Chicagoans also continued to hold a rather benign impression of smoke. Perhaps most important, however, the men who had organized the Society had done so with the protection of the 1893 World's Fair's White City as its primary goal. With the World's Fair successfully completed, undoubtedly the most active members in the Society lost their compelling interest in smoke abatement (Rosen, "Businessmen against Pollution in Late-Nineteenth-Century Chicago," 380–85).

39. For a description of the New York City environment in the late 1800s see John Duffy, *A History of Public Health in New York City, 1866–1966* (New York: Russell Sage Foundation, 1974), 112–42. For Cincinnati see Zane Miller, *Boss Cox's Cincinnati* (Chicago: University of Chicago Press, 1968), 3–24.

40. Both the Optimist Club and the Manufacturer's Club organized antismoke committees and discussed the smoke nuisance in the 1890s, but neither organization had initiated an organized campaign to improve municipal regulation. See *Smoke Prevention and Fuel Gas Considered by the Manufacturer's Club* (Cincinnati, 1896); Samuel Skinner, "Smoke Abatement."

41. Most of the Woman's Club's members lived in the suburbs that sat upon the hills north of the central city basin. Woman's Club, "Reports of the Department of Civics, 1895–1906," 186, 188; Woman's Club, "Minutes of the Executive Board, 1902–," 130–31, 136. The Woman's Club's records are privately held by the club.

42. Reed enjoyed national recognition among physicians, serving as president of the American Medical Association and, at the time of the address to the Woman's Club, chairman of the legislative committee of the A.M.A. Twice summaries of Reed's speech appeared in the *Journal of American Medical Association* (44 [1905]: 1619–20; 49 [1907]: 813), and *American Medicine* reprinted the entire speech (9 [1905], 703). The *Medical Record* also reported on Reed's address and noted that the *St. Louis Medical Review* had published the speech (*Medical Record* 67 [1905]: 860). Cleveland's Smoke Inspector John Krause extensively quoted Reed in a speech he gave before the Engineers' Society of Western Pennsylvania (*Proceedings of the Engineers' Society of Western Pennsylvania* 24 [1908]: 99). Reed's address also circulated in pamphlet form and is preserved at the Cincinnati Historical Society (Charles A. L. Reed, "An Address on the Smoke Problem," delivered before the Woman's Club of Cincinnati, 24 April 1905, 1–4).

43. Woman's Club, "Reports of the Department of Civics," 201, 205; Woman's Club, "Annual Report for the Year Ending June 5, 1905," 11. The annual report issued in 1905 included a map of several Cincinnati blocks which indicated the level of smoke issued from the various buildings in that section of the city. The accompanying text noted that smokeless chimneys stood as proof that smoke could be prevented. See also Charles Reed, "The Smoke Campaign in Cincinnati," remarks made before the National Association of Stationary Engineers, Cincinnati, 10 July 1906, 6.

44. The Smoke Abatement League published a list of members with each annual report. For brief biographies of Cincinnati's prominent citizens see Charles Frederick Goss, *Cincinnati: The Queen City, 1788–1912* (Cincinnati, 1912).

45. Cincinnati *Times-Star*, 11 January 1910.

46. As was commonplace, Reed used "fireman" to describe those men who tended and fed fires.

47. Reed, "The Smoke Campaign in Cincinnati," 5.

48. Editorial, New York *Times*, 27 April 1905. The next year Barney noted that the Board of Health's antismoke activity had decreased dramatically over the previous three years. In 1902 the city had investigated 714 cases and prosecuted 57, but in 1905 those figures dropped to 117 and 1, respectively. Obviously the Board of Health had become incapable of stopping the growing smoke nuisance (New York *Times*, 6 March 1906).

49. New York *Times*, 5 September 1905; 8, 15 March and 4, 17 May 1906; *Industrial World* 41, pt. 2 (1907): 855–57, indicated as V–VII in the volume; *Medical Record* 70 (1906): 420. Most of the smoke convictions in 1906 resulted in suspended sentences, due, at least according to the presiding justice, to a scarcity of hard coal. See New York *Times*, 18 May 1906. The New York Board of Health *Annual Report* for 1906 listed different, but similar numbers for arrests and convictions during the year, with 165 of 211 ending in conviction and 17 still pending at the time of the report. With the great number of suspended sentences, however, the city had collected only $240 in fines from five convictions. See *Annual Report, 1906*, 130–31.

50. *Industrial World* 41, pt. 2 (1907): 856–57; Chicago *Record-Herald*, 22 September 1906.

51. New York *Times*, 6 March 1906, 27 June 1907, 13 April 1907.

52. A. A. Straub, "Some Engineering Phases of Pittsburgh's Smoke Problem," *Mellon Institute Smoke Investigation Bulletin No. 8* (1914), 12. The Mellon Institute collected copies of the antismoke ordinances of seventy-five American cities by 1915. See "Proceedings of the Tenth Convention of the International Association for the Prevention of Smoke," Cincinnati, 1915, 28.

53. The official charged with protecting Philadelphia from nuisances, Board of Health President William Ford, agreed that his was "not a very smoky city," but he urged action before the nuisance became entrenched (*Franklin Institute* 144 [July, 1897]: 36, 32).

54. Two other aspects of progressive-era politics deserve some attention, since they aided in the rapid expansion of interest in environmental reform. First, urban rivalry, although not peculiar to the progressive era, became particularly intense as cities vied with one another to attract capital, labor, and transportation routes. In an effort to improve their cities' positions in the great municipal race for growth, urban boosters undertook a myriad of reforms to make their cities more attractive to investors and laborers. Smoke abatement often played a part in this reformism, and antismoke crusaders around the nation argued that pure air would give their cities an advantage in the competition for economic growth, not simply because the quality of the air was important, but because the effort to clear the air said much about a city and its citizens. As Charles Reed asked of Cincinnati as he launched the Smoke Abatement League, "Shall it be said that we have less genuine enterprise, less civic pride, less public decency than other cities?" (Reed, "The Smoke Campaign in Cincinnati," 8). See William Cronon, *Nature's Metropolis* (New York: Norton, 1991), 31–46, for a discussion of civic boosters.

Second, the progressive era witnessed a significant change in municipal politics. Interest groups evolved into dominant organizations in the creation, expression, and implementation of urban policies. The development of interest group politics gave new or louder voices to several environment-conscious segments in society, including middle-class women and civic boosters, who kept issues like smoke abatement before the public. For the importance of organization and the development of interest groups see Samuel Hays, *The Response to Industrialism* (Chicago: University of Chicago Press, 1957), 48–70.

55. The phrase "environmental amenities" comes from Samuel Hays, *Beauty, Health, and Permanence* (New York: Cambridge University Press, 1987), 22. See Martin Melosi, "Environmental Crisis in the City: The Relationship between Industrialization and Urban Pollution," in *Pollution and Reform in American Cities*, ed. Melosi, 3–31.

56. Reed, "The Smoke Campaign in Cincinnati," 4.

57. While antismoke reformers rarely offered broad critiques of the industrial economy, choosing instead to attack its many problems piecemeal, activists well un-

derstood that smoke was an industrial problem. Most reformers, however, chose to view smoke and other environmental problems as signs of individual failures, rather than of systemic defects. Reformers most often blamed individual firemen and engineers for incompetence in creating smoke, but they also blamed proprietors. In 1904, the Chicago *Record-Herald* blamed a few thousand proprietors for lining the lungs of Chicago's two million residents with soot. "The few thousand," the editorial read, "think they make a little extra profit and the two million suffer grievously in health and comfort" (Chicago *Record-Herald*, 10 December 1904). In 1906, the *Medical Record* begged for action in New York, "unless we are content to sacrifice the beauty of our Italian sky to the greed of wealthy corporations" ("The Smoke Nuisance," *Medical Record* 69 [1906]: 392).

58. Cincinnati's Edward Jerome, superintendent of the Smoke Abatement League, was just one of many observers to note the contradiction inherent in smoke clouds hanging over the world's most civilized cities. "Are we not living in the most enlightened age of the world?" he asked. "If so, let us prove it to ourselves and to others" (Smoke Abatement League, "Annual Report, 1911," 10).

59. "The Smoke Nuisance," *Sanitary News* 18 (1891): 129. Barbarism quote from editorial, New York *Tribune*, 6 March 1898. The Geological Survey's H. M. Wilson also declared in 1909, "The smoky city is to be a sign and relic of barbarism" (*Outlook* 93 [1909]: 6); Cincinnati *Times Star*, 1 June 1908. William T. Stead, in his popular 1894 work *If Christ Came to Chicago*, expressed concern over the state of civilization in that city, often using the poor state of the environment as evidence of degeneracy. The argument that smoke threatened civilized society also found resonance in Britain. As Alfred R. Wallace pleaded to his fellow Brits, "We claim to be a people of high civilisation, of advanced science, of great humanity, of enormous wealth! For very shame do not let us say 'We *cannot* arrange matters so that our people may all breathe unpolluted, unpoisoned air!'" (Alfred R. Wallace, *Man's Place in the Universe* [New York: McClure, Phillips, 1903], 257).

60. This should not suggest, however, that individual polluters never took steps to control their emissions without government involvement. In one important example, in the years around 1900 the Jones and Laughlin Steel Company built its Hazelwood beehive coke ovens with tall stacks to disperse emissions, thus relieving nearby residents of some of the pollution (Joel Tarr, "Searching for a 'Sink' for an Industrial Waste," in *The Search for the Ultimate Sink* [Akron: University of Akron Press, 1996], 385–411).

61. See Delos Wilcox, *The American City* (New York: Macmillan, 1904), 200–202, for a discussion of civic cooperation vis-à-vis the urban environment.

four The Atmosphere Will Be Regulated

1. See Stanley Schultz, *Constructing Urban Culture* (Philadelphia: Temple University Press, 1989), 42–47, 50–52, for a discussion of the changing role of nuisance law in

burgeoning antebellum cities. For a brief and remarkably insightful interpretation of nineteenth-century law see James Willard Hurst, *Law and the Conditions of Freedom* (Madison: University of Wisconsin Press, 1965). The best single source on the development of nuisance law in the nineteenth century remains Horace Gay Wood, *A Practical Treatise on the Law of Nuisances in Their Various Forms* (Albany: John D. Parsons, 1875).

2. The annual reports of municipal health departments generally contain a list of nuisances reported and abated. This list is taken from the Cincinnati Board of Health *Annual Report*, 1869–1899, and its composition varied little over the last three decades of the 1800s.

3. *Journal of the Franklin Institute* 144 (1897): 52–61. Other cities, including St. Louis and Pittsburgh, gave the power to enforce their antismoke laws to other departments, such as the departments of public safety, public improvements, or public works. This placement did not necessarily require that smoke abatement officials know any more about smoke creation and prevention than health officials, however.

4. *A Digest of the Act of Assembly Relating to and the General Ordinances of the City of Pittsburgh, from 1804–1908* (1909), 568; Milwaukee Board of Health, *Annual Report* (1899), 44; *General Ordinances and Resolutions of the City of Cincinnati in Force April, 1887* (1887), 269.

5. Bryan was an active member of the American Society of Mechanical Engineers and the Engineers' Club of St. Louis. See William H. Bryan, "The Problem of Smoke Abatement," An Address Delivered before the Optimist Club of Cincinnati, 20 May 1899, 28. Of course Bryan had incentive to exaggerate the success of his department, but while his estimate of the smoke reduction may have been far too high, clearly the enforcement of the smoke law had a positive effect on the air quality of St. Louis. See also Bryan, "Smoke Abatement in St. Louis," *Journal of the Association of Engineering Societies* 27 (December 1901): 215–19. For the ordinance see Lucius Cannon, *Smoke Abatement: A Study of Police Power* (St. Louis: St. Louis Public Library, 1924), 212.

6. In this sense common law treated the smoke nuisance differently from many other nuisances. For example, health officials could force a property owner to drain standing water from his land without having to prove that the water injured or annoyed other citizens. The same held true for garbage and odor nuisances. These were all nuisances per se, and courts assumed their threat to human health.

7. Cannon, *Smoke Abatement: A Study of Police Power*, 214–15; St. Louis *Post-Dispatch*, 17 November 1897. It was not unusual for firms to join forces in challenging antismoke ordinances in court. Just as Heitzeberg received support from other St. Louis companies, so too did the Springfield Gas Light Company as it challenged the Springfield, Massachusetts law in 1906. The gas company received support from a local brewery and two major railroads, the New York Central and the New York, New Haven, and Hartford (*Municipal Journal and Engineer* 20 [1906], 583).

8. Bryan, "Smoke Abatement in St. Louis," 28. The 1899 law was constitutional since it treated smoke like any other nuisance. The city was empowered to prosecute residents who used their property in a manner that harmed neighbors.

9. Ibid., 26.

10. For a discussion of court challenges to the Cleveland and St. Paul antismoke ordinances see Cannon, *Smoke Abatement: A Study of Police Power*, 245–47, 280–81. Cleveland Chamber of Commerce, "Report of the Committee on Smoke Prevention, 15 May 1912," 4; Smoke Abatement League of Hamilton County, "Annual Report" (1911), 6. On state control of municipalities see Schultz, *Constructing Urban Culture*, 66–75, and Jon Teaford, *The Unheralded Triumph* (Baltimore: Johns Hopkins University Press, 1984), 83–102.

11. *Moses v. United States*, 16 D.C. App. 428 (1900). Although this case involved the special relationship between Congress and the capital district, the decision made clear its relevance for cases involving municipal regulation in the states. The court summarized, "The power of Congress to enact regulations affecting the public peace, morals, safety, health and comfort, within the District of Columbia, is the same as that of the several State legislatures within their respective territorial limits."

12. *State v. Tower*, 185 MO 79 (1904). See also *Glucose Refining Co. v. City of Chicago*, 138 Fed 209 (1905); *Bowers v. City of Indianapolis*, 169 Ind 105 (1907).

13. *Northwestern Laundry v. City of Des Moines*, 239 U.S. 486, 491–92 (1916). The judicial approval of municipal regulation of smoke emissions should not suggest that all convicted smoke offenders lost their appeals. In New York, for example, the New York Central Railroad won an appeal which argued that the city had failed to consider the limitations of technology when issuing fines. The railroad successfully argued that locomotive fires must smoke when first ignited (*People v. New York Central & H. R. R. R. Co.*, 159 NY App. 329 [1913]).

14. Chicago *Record-Herald*, 25 April 1904 and 4, 10 October 1904. For judicial support of Chicago smoke regulation see *Harmon v. Chicago*, 110 Ill. 400 (1884), and *Field v. Chicago*, 44 Ill. App. 410 (1892).

15. Chicago *Record-Herald*, 10, 11 October 1904.

16. Chicago *Record-Herald*, 6 February 1905.

17. Chicago *Record-Herald*, 15 February 1905.

18. Chicago *Record-Herald*, 17 February 1905 and 24 May 1905; *Glucose Refining Co. v. City of Chicago*, 138 Fed 215 (1905).

19. New York *Times*, 9 January 1906; Chicago *Record-Herald*, 19, 26 July 1906; 16 August 1906; 13 September 1906 (for Schubert quote); 20, 27 September 1906.

20. Chicago *Record-Herald*, 2 October and 24 November 1906; Jacob Zeller, "The Corn Products Refining Co.," *Moody's Magazine* 15 (1913): 399–405.

21. Chicago *Record-Herald*, 30 January 1906.

22. Ibid.

23. "Soot? Enginemen Pay," Chicago *Record-Herald*, 15 February 1906. In seeking refuge in jury trials the Illinois Central adopted a strategy that had proven successful in the past. In 1893, a series of hung juries, small fines, and even acquittals had helped smoke offenders beat the 1881 antismoke law and destroy the Society for the Prevention of Smoke (Christine Meisner Rosen, "Businessmen against Pollution in Late-Nineteenth-Century Chicago," *Business History Review* 69 [1995]: 377–81).

24. Chicago *Record-Herald*, 6 September 1906.

25. Railroad lawyers did continued to use the tactic, however. The Pennsylvania Railroad's General Agent and Superintendent W. H. Scriven advised his company's lawyers to pursue jury trials in 1911: "I believe that we would secure better treatment from a jury, as it would probably have on it at least one or two railroad men, coal men, or other fair minded persons." The solicitors followed his instructions and demanded jury trials for twenty-eight pending cases. The city responded by immediately settling eleven cases for a total of $100, an indication that they thought them unwinnable before a jury. Indeed, the city proved so unwilling to take any cases before a jury that the Pennsylvania Railroad secured a settlement of $125 plus court costs for forty-three violations pending in April 1911 (Scriven to Loesch, Scofield, and Loesch, 28 February 1911; Loesch, Scofield, Loesch to Scriven, 10 March, 21 April 1911, Pennsylvania Railroad MSS, box 1271, folder 16, Hagley Library).

26. Chicago *Record-Herald*, 16 March, 18 December 1906.

27. Milwaukee, "Common Council Action on Smoke Control, 1896–1947," 1–4; Milwaukee Board of Health, *Annual Report* (1899), 44. Although the Milwaukee ordinance was typical in many respects, Wisconsin law ensured that it would have one peculiarity. In 1911, the state legislature of Wisconsin had passed a new enabling law for Milwaukee which "authorized and empowered" the city "to regulate and prohibit the emission of dense smoke into the open air within the corporation limits" and "within a distance of one mile therefrom." The 1914 legislation reflected this change, as Milwaukee's bureau of smoke suppression regulated emission in and around the city. Wisconsin had recognized in law that smoke traveled across municipal boundaries (Bureau of Smoke Suppression, City of Milwaukee, "Smoke Suppression Ordinances" [1914], 3–9).

28. Pittsburgh Bureau of Smoke Regulation, *Handbook* 1918, 7. The chart bore the name of the man who developed the scale in Paris in the late 1890s, Max Ringelmann. Use of the chart spread quickly after its introduction to the United States, particularly through the American Society of Mechanical Engineers, which presented the smoke chart to the United States in late 1899. See Lester Breckenridge, "How to Burn Illinois Coal without Smoke," *University of Illinois Engineering Experiment Station Bulletin Number 15* (1907), 11–13; A.S.M.E. *Transactions* 21 (1900): 97–99; "Ringelmann Smoke Chart," Bureau of Mines, *Information Circular 8333* (1967). Some cities used a different device to judge the density of smoke: an umbrascope, a tube with grey-tinted glasses placed together. If smoke was visible through the darkened scope, then the stack violated the ordinance. Cincinnati, Detroit, Akron, and several other cities opted for the umbrascope over the Ringelmann Chart (J. F. Barkley, "Some Fundamentals of Smoke Abatement," Bureau of Mines, *Information Circular 7090* [1939], 19–20).

29. New York *Times*, 5 September 1911. For the ordinance see *People v. New York Edison Co.* 159 NY App. 787 (1914). This ordinance was innovative in one area: its inclusion of automobile exhaust as a nuisance. The health department did seek arrests of operators of smoky autos.

30. New York *Times*, 19 July 1913.

31. New York *Times,* 21 July 1913; 27 July 1913. As the *Times* attempted to save the city's smoke ordinance, others attempted to save Yosemite's Hetch Hetchy Valley. On July 21, the editorial supporting the antismoke ordinance ran next to a lengthy letter from John Muir concerning "Hetch Hetchy Invaders," while the July 27 antismoke editorial ran next to a letter from Robert Underwood Johnson decrying the plan put forth by the city of San Francisco to build a dam within the national park.

32. "New York Smoke Ordinance Unconstitutional," *Power* 38 (1913): 174. *Power* took a particular interest in the smoke issue in the 1910s, publishing frequent reports concerning antismoke engineering advances and numberless editorials on the subject. Not coincidentally, Osborn Monnett, one of Chicago's smoke inspectors and a nationally known antismoke activist, served as the magazine's western editor.

33. *People v. New York Edison,* 159 NY App. 789 (1914).

34. *Industrial World* 48, pt. 1 (1914): 137. At the time of the initial antismoke activism in Youngstown, it was a booming small city of about eighty thousand residents. Some members of the Chamber of Commerce expressed anxiety about any ordinance that would regulate emissions, including T. J. Bray, the president of the Republic Iron and Steel Company, who suggested that rigid legislation would force some businesses to close due to limited capital for plant improvements.

35. *Industrial World* 48, pt. 1 (1914): 135. Smoke abatement even gained attention in Baltimore's cleaner neighbor, Washington. Congress declared thick black or gray smoke a nuisance in the District of Columbia and prohibited its emission in 1899, setting penalties from ten to one hundred dollars for polluters. Washington was not a major manufacturing city, but many residents in the capital complained about the effect of smoke from office buildings on the appearance of the city's many white facades. In 1905, after the ordinance had brought only slow progress in abating dense smoke, one resident, Theodore Roosevelt, took his complaint directly to Congress. In his State of the Union address, the president recommended that the ordinance "be made more stringent by increasing both the minimum and maximum fine; by providing for imprisonment in cases of repeated violation; and by affording the remedy of injunction against the continuation of the operation of plants which are persistent offenders." Roosevelt also recommended that Congress add more inspectors to the smoke department so that more offenders might be brought to court (Cannon, *Smoke Abatement: A Study of Police Power,* 285; *Congressional Record,* 95th Cong., 1st sess., 1905, 40: 102).

36. In 1913, for example, Lukens counted 207 nuisances abated for the year, 71 of which had turned to hard coal, and another 44 had begun mixing hard and soft coal to reduce smoke. See *Annual Report of the Mayor of Philadelphia* (title varies with name of the mayor), 1905–1913. Philadelphia was not the only city in which substitution of smokeless fuels became the most common solution to smoking stacks. In Grand Rapids, for example, most improvements in stack performance in the early 1910s involved switching to smokeless coals or coke (*Industrial World* 47, pt. 1 [1913]: 147–48).

37. Formed in 1895, the Civic Club was in part the descendant of the Ladies' Health Protective Association. Unlike the association, however, the Civic Club included both

men and women, with the men holding a preponderance of the official posts (Angela Gugliotta, "Women and Anti-Smoke Activism in Pittsburgh, 1880–1920" [paper delivered before the biennial meeting of the American Society for Environmental History, Baltimore, Md., March 1997]).

38. Although the Chamber of Commerce was instrumental in the creation of the Smoke and Dust Abatement League, at least one prominent member, the University of Pittsburgh's John O'Connor, regretted the influence of the Chamber over the League. O'Connor, who hoped the league would become a bulwark for pure air, wrote to fellow member and fellow economics professor J. T. Holdsworth to complain about the performance of William Todd, the Chamber of Commerce member who served as the first president of the league. "You know well the reputation the Chamber of Commerce has," wrote O'Connor, "for devitalizing and discouraging anything that bears even a slight resemblance to an active movement." O'Connor asked that Holdsworth vote against the reelection of Todd. At the next meeting the league elected A. A. Hamerschlag, director of the Carnegie Institute of Technology, president, and O'Connor became secretary (O'Connor to Holdsworth, 22 November 1913, Smoke Investigation Activities MSS, ser. 1, folder 2, University of Pittsburgh Library; *Industrial World* 48, pt. 1 [1914]: iv).

39. *Industrial World* 45, pt. 2 (1911): 841, 1010, 1154; 46, pt. 2 (1912): 996; "The Smoke and Dust Abatement League," 1916, a brief pamphlet published in Pittsburgh. Henderson's estimates in the decrease in smoke relied on U.S. Weather Bureau reports of smoky days. In the first six months of 1912, Pittsburgh saw a total of ninety-three smoky days, while in the same period in 1916 it saw only fifty. Of course these statistics could not be wholly scientific, but they could reveal a trend (International Association for the Prevention of Smoke, "Proceedings of the Eleventh Annual Convention" [1916], St. Louis, 94–95).

40. Rochester Chamber of Commerce, "The Smoke Shroud: How to Banish It" (191?), 18–19.

41. "Smokeless Cities of To-Day," *Harper's Weekly* 51 (1907): 1139. This altered perception of smoke was certainly more common among the middle class. For the continuing acceptance of the perception of smoke as a symbol of prosperity by the working class, even into the present, see Joel Tarr and Bill C. Lamperes, "Changing Fuel Use Behavior and Energy Transitions: The Pittsburgh Smoke Control Movement, 1940–1950," *Journal of Social History* 14 (1981): 563.

42. Chicago Association of Commerce, *Smoke Abatement and Electrification of Railway Terminals in Chicago* (1915), 173.

43. Engineering and industrial trade journals were replete with articles concerning the latest advances in smoke abatement devices. See particularly *Iron Age, Industrial World, Power, Engineering News,* and *Factory.* For a brief discussion of the four major categories of smoke prevention devices see William Bryan, "Smoke Abatement," *Cassier's Magazine* 19 (1900): 22–24.

44. *Iron Age* 90 (1912): 19. *Iron Age* reprinted Warner's letter in full, vol. 80 (1907):

1149. See Harold C. Livesay and Glenn Porter, "William Savery and the Wonderful Parsons Smoke-Eating Machine," *Delaware History* 14 (1971): 161–76. After relating the failed efforts of this inventor, Livesay and Porter conclude that the battle against smoke was "lost on the practical grounds of technology and economics, for entrepreneurship failed at the critical time to produce a device that was workable and cheap." This conclusion is not supported by the evidence. Although no single device provided an absolute cure for smoke, many capable devices and designs did effectively reduce smoke. As for cheapness, a relative term, city residents, both those who produced smoke and those who lived in it, had to determine what price clear skies warranted.

45. Chicago *Record-Herald*, 4 April 1904; 23 August 1908.

46. Milwaukee, *Annual Report of the Smoke Inspector* (1912), 14; Alburto Bement, "The Suppression of Industrial Smoke with Particular Reference to Steam Boilers," *Journal of the Western Society of Engineers* 11 (1906): 722. Ordinances requiring permits for new furnace and boiler construction usually explicitly excluded residences, except large apartment buildings.

47. John W. Krause, "Smoke Prevention," *Proceedings of the Engineers Society of Western Pennsylvania* 24 (1908): 101. See the discussion that followed Alburto Bement's lecture to the Western Society of Engineers (*Journal of the Western Society of Engineers* 11 [1906]: 703–4 and passim). Those with an ultimate faith in science and technology tended to place heavy blame on the incompetence of firemen and engineers throughout the early 1900s, allowing them to remain optimistic about the technological future and pessimistic about the fate of the ignorant and dirty poorer classes, particularly immigrants.

48. *Locomotive Firemen's Magazine* 27 (1899): 61.

49. Charles Poethke, *Annual Report of the Smoke Inspector*, Milwaukee, for the year 1909, p. 7; for the year 1912, p. 9.

50. Cincinnati *Times-Star*, 4 April, 3 March 1913; Cincinnati *Commercial Tribune*, 8 April 1913. Undoubtedly the smoke from steamships constituted only a small fraction of city air pollution. In Chicago, for example, even with its busy Lake Michigan port, one estimate placed the contribution of steam vessels at less than 1 percent of all smoke in the city. See charts in Goss, *Smoke Abatement and Electrification of Railway Terminals in Chicago* (Chicago, 1915), facing p. 178.

51. New York *Tribune*, 13 September 1899; House Committee on Interstate and Foreign Commerce, *Supervisor of New York Harbor, Etc.*, 56th Cong., 1st sess., 1900, H. Rept. 478, 1–5.

52. New York *Tribune*, 28 February 1900; 56th Cong., 1st sess., 1900, H. Rept. 478, 4–7. The 56th Congress had the best chance for passage of this bill, as New York Representative Nicholas Muller sat on the Interstate Commerce Committee. The federal legislation concerning New York Harbor which passed after Muller's resignation in 1901 simply allotted more funding for lighthouses and did not address the regulation of smoke. Although Congress did not pass legislation concerning air pollution outside the District of Columbia until 1955, the Supreme Court did rule on interstate air

pollution as early as 1907. In that year the court ruled in favor of the state of Georgia, which had sued for an injunction against two smelting companies in Tennessee. Writing for the majority, Oliver Wendell Holmes argued that the state "has the last word as to whether its mountains shall be stripped of their forests and its inhabitants shall breathe pure air." In essence, Holmes declared the pollution that crossed the state boundary had interfered with the state of Georgia's right to regulate its own environment. See *Georgia v. Tennessee Copper Company*, 206 U.S. 230. For early congressional discussion of federal involvement in air pollution control see *Congressional Record*, 84th Cong., 1st sess., 1955, 101, pt. 6: 7248–50, and pt. 8: 9923–25.

53. New York *Times*, 26, 27, 28, 30 March 1913.

54. Cincinnati *Times-Star*, 4 April 1913; Cincinnati *Enquirer*, 6 April 1913. Although More claimed that the absence of locomotives and steamboats had caused the smoke abatement, reports concerning flood damage contained frequent references to closed factories, which also contributed to the cleaner air. See New York *Times*, 30 March 1913; Cincinnati *Enquirer*, 5 April 1913.

five Engineers and Efficiency

1. Charles Mulford Robinson, *The Improvement of Towns and Cities* (New York: Putnam's, 1901), 61. Another lay reformer, City Club of Chicago's Robert Kuss, recognized that "unless engineering thought comes to the rescue, the law itself may become inoperative because of impracticability" (*Journal of the Western Society of Engineers* 11 [1906]: 697). *Sanitary News* 18 (1891): 53, 388; Ernest L. Ohle and Leroy McMaster, "Soot-Fall Studies in Saint Louis," *Washington University Studies* 5 (July 1917): 3–8.

2. Charles H. Benjamin, "Smoke and Its Abatement," *Transactions of the American Society of Mechanical Engineers* 26 (1905): 743. Some engineering trade magazines could also wax eloquent on the peril of smoke. *Power and the Engineer* (32 [1910]: 550) editorialized: "And let us not forget that after all, this is as much a problem in ethics as in engineering; that people are beginning to wake up to their rights; that everyone has as much title to pure air for breathing as to pure water for drinking, and finally, that the man who allows black smoke and soot to escape from his chimneys is not only a poor manager but an undesirable citizen."

3. *Journal of the Western Society of Engineers* 11 (1906): 731.

4. Certainly these two poles of argumentation, smoke abatement as an efficiency issue versus a health, beauty, and cleanliness issue, are not antithetical. Most anti-smoke activists, both lay and expert, argued that smoke was dirty *and* inefficient, that it threatened health *and* wasted coal, that it was immoral *and* uneconomical. But the gradual shift from one pole toward the other did have significant consequences for municipalities and their residents, as the following chapters reveal.

5. Engineers came to dominate efforts to solve other important environmental problems in the early 1900s, including water pollution, garbage collection, and sewage removal. See Joel Tarr, "Searching for a 'Sink' for an Industrial Waste: Iron-

Making Fuels and the Environment," *Environmental History Review* 18 (1994): 24; Martin Melosi, *Garbage in the Cities* (College Station: Texas A & M University Press, 1981), 79–104; and Stanley Schultz and Clay McShane, "To Engineer the Metropolis: Sewers, Sanitation, and City Planning in Late-Nineteenth-Century America," *Journal of American History* 65 (1978): 389–411.

6. William Goss, "Smoke Responsibility Cannot Be Individualized," *Steel and Iron* 49, pt. 1 (1915): 226. Engineers had long played important roles in building industrial cities, particularly through the construction of sewers and water delivery systems, but in the increasingly complex cities of the turn of the century and with the growing authority of municipalities, engineers became ever more important to cities and their governments. For a discussion of the growing importance of engineers in the late 1800s, see Stanley Schultz, *Constructing Urban Culture* (Philadelphia: Temple University Press, 1989), 153–205; Jon Teaford, *The Unheralded Triumph* (Baltimore: Johns Hopkins University Press, 1984), 133–41.

7. Edwin T. Layton Jr., *The Revolt of the Engineers* (Cleveland: Case Western Reserve University Press, 1971), 3, 53–68.

8. See Stanley Schultz and Clay McShane, "Pollution and Political Reform in Urban America: The Role of Municipal Engineers, 1840–1920," in *Pollution and Reform in American Cities, 1870–1930*, ed. Martin Melosi (Austin: University of Texas Press, 1980), 155–72; Schultz, *Constructing Urban Culture*, 183–205; Layton, *The Revolt of the Engineers*, 13–14. On the efficiency craze see Samuel Haber, *Efficiency and Uplift* (Chicago: University of Chicago Press, 1964), 51–74. Public interest in efficiency exploded in 1911, after Louis Brandies referred to scientific management during a railroad rate increase inquiry before the Interstate Commerce Commission in late 1910. Brandies, arguing against the rate hike, insisted that railroads could raise wages and cut rates if they would embrace a system of scientific management. Many Americans quickly understood the power of a new system of management which promised better results for business, labor, and consumers. Efficiency through more scientific organization and operation, proponents argued in the years that followed, would greatly augment production and profit, for all of society. See Horace Drury, *Scientific Management* (New York: Longmans, Green, 1922), 35–48.

9. "Report of the Special Committee on Prevention of Smoke," *Journal of the Association of Engineering Societies* 11 (1892): 291–327; "The Smoke Nuisance and Its Regulation, with Special Reference to the Condition Prevailing in Philadelphia," *Journal of the Franklin Institute* 143 (1897): 393–424, and 144 (1897): 17–61.

10. "Abatement of the Smoke Nuisance," *Journal of the Association of Engineering Societies* 30 (1903): 41–45. American engineers often referred to the work of British, German, or other European engineers, many of whom had long been engaged in smoke abatement research. As early as 1856, Philadelphia's *Journal of the Franklin Institute* reprinted a smoke article from Britain's *Mechanics' Magazine* (61 [1856]: 67, 113).

11. Although many observers referred to the work of Franklin as the birth of antismoke engineering, he was not the first to create a down-draft furnace nor the first to

take up the smoke issue while designing furnace equipment. The Scottish inventor James Watt—also mentioned frequently by engineers interested in designing smokeless equipment—invented the down-draft furnace, which pulled air down through the hot coals and then up past the boiler. This method ensured more complete mixing of available oxygen with the fire, but it did not ensure smokelessness. See *Engineers' Society of Western Pennsylvania* 8 (1892): 301; Leonard W. Labaree, ed., *The Papers of Benjamin Franklin*, vol. 13 (New Haven: Yale University Press, 1969), 197.

12. D. H. Williams and R. B. Fitts, "D. H. Williams' Improved Apparatus for the Combustion of Smoke in Steam Boiler Furnaces," 1860; Samuel Kneeland, M.D., "On the Economy of Fuel, and the Consumption of Smoke, as Effected by Amory's Improved Patent Furnace," 1866; D. G. Power, "A Treatise on Smoke: Its Formation and Prevention," 1879.

13. This type of public recommendation differed from written testimonials delivered by salesmen, which prospective buyers often ignored. For the unwillingness of railroads to accept testimonials as evidence of the effectiveness of smoke abatement devices see Harold C. Livesay and Glenn Porter, "William Savery and the Wonderful Parsons Smoke-Eating Machine," *Delaware History* 14 (1971): 171.

14. Robert Moore, "Smoke Prevention," *Journal of the Association of Engineering Societies* 8 (1889): 201–4; *Industrial World* 48, pt. 1 (1914): 134.

15. Association of the Transportation Officers of the Pennsylvania Railroad, "Report of the Committee on Motive Power," 21 November 1894, Pennsylvania Railroad MSS, box 410, folder 14, Hagley Library, Wilmington, Delaware. Some engineers continued to identify anthracite, or a semibituminous coal, as the only true solution to the smoke problem. In 1903, Charles Pugh, Second Vice President of the Pennsylvania, wrote to General Superintendent of Motive Power A. W. Gibbs concerning "a rather formidable effort" made in Philadelphia's council to pass "impracticable and onerous" antismoke legislation. Pugh asked Gibbs to find some way to "reduce the nuisance to the lowest limit possible." Gibbs replied that anthracite provided the surest solution, and that where that proved impractical semibituminous Pocahontas coal might be acceptable (Pugh to Gibbs, 9 September 1903; Gibbs to Pugh, 11 September 1903, PA RR MSS, box 703, folder 1, Hagley Library).

16. Altoona Railroad Club of the Pennsylvania Railroad, "Report of Maintenance of Equipment Committee" (1910), 1–2; Livesay and Porter, "William Savery and the Wonderful Parsons Smoke-Eating Machine," 170. For Pennsylvania Railroad research see David Crawford in *Industrial World* 47, pt. 2 (1913): 1097–98. Crawford to Peck, 29 December 1910, PA RR MSS, box 1291, folder 15, Hagley Library. For Searle's comments on the Crawford Stoker see Pittsburgh Health Bureau, *Annual Report* (1911), 368. For locomotive inventory see Pennsylvania Railroad Company, *67th Annual Report* (1913), 66.

17. For the steam jet tests the Altoona shops used a specially designed test locomotive, held stationary by a dynamometer, which measured its power, and resting upon wheels which operators could brake to simulate varying loads. The engineers

used the Ringelmann Chart to determine that steam jets could reduce smoke. See D. F. Crawford, "The Abatement of Locomotive Smoke," *Industrial World* 47, pt. 2 (1913): 1096–97. The railroad had previously used this locomotive to test experimental fuel as well as equipment. In 1908, the Geological Survey reported on the progress of research conducted at the Altoona shops using government-made coal briquettes and the railroad's test locomotive. See William F. M. Goss, "Comparative Tests of Run-of-Mine and Briquetted Coal on Locomotives," *Geological Survey Bulletin 363* (1908).

18. "Report of Test Made at the Altoona Testing Plant for General Manager's Association," 1913; Chicago General Manager's Association Report, 15 May 1913; Chicago Gen. Man. Ass. Circular No. 740, 29 July 1913, PA RR MSS, box 1272, folder 2, Hagley Library. Crawford delivered a lengthy speech on his work to the International Association for the Prevention of Smoke in Pittsburgh in an effort to convince smoke abatement officials that the railroads understood the need to reduce their emissions and that they had invested large amounts of money in that endeavor. Crawford undoubtedly had some effect on city officials. See *Industrial World* 47, pt. 1 (1913): 130–31.

19. *Engineering News* 35 (1896): 9; *Iron Age* 66 (1900): 25.

20. In response to the *Engineering News* article on smoke prevention, one employee of the Chicago, Burlington, and Quincy Railroad replied in part: "There is no 'cure all' for the varied ills that boiler and furnace plants are heir to, but an intelligent study of each case will always show a chance for improvement and in a majority of cases will bring about comparatively complete results" (*Engineering News* 35 [1896]: 93).

21. Bird quote comes from his talk, "City Supervision of New Boiler Plants," found in "Proceedings of the International Association for the Prevention of Smoke, 3rd Annual Convention, 1908," 48.

22. The Pennsylvania Railroad's research concerning the smokeless consumption of soft coal on locomotives dated back to 1859, when the company hoped to switch its passenger trains from wood fuel to coal. With most existing technology, however, dense smoke from soft coal locomotives made passengers uncomfortable. After a series of test runs out of Altoona, the railroad found a combination of equipment which allowed them to expand their use of inexpensive bituminous coal. See *Railroad Gazette* 43 (1907): 719–24.

23. A. S. Vogt to A. W. Gibbs, 31 July 1903, PA RR MSS, box 703, folder 1, Hagley Library.

24. G. E. Rhoades to E. D. Nelson, 1 September 1906, PA RR MSS, box 703, folder 1, Hagley Library. The sample apparently never arrived, and at least one subsequent attempt to locate Mulvaney failed.

25. E. R. Dunham to A. W. Gibbs, 18 February 1909; G. B. Koch to E. D. Nelson, 9 March 1909, PA RR MSS, box 703, folder 1, Hagley Library.

26. D. M. Perine to A. W. Gibbs, 8 November 1907; Gibbs to Perine, 11 November 1907; Koch, Bodson, Sproul, "Smoke Prevention—Visit to Chicago," 22 November 1907, all in PA RR MSS, box 703, folder 1, Hagley Library.

27. As late as 1912 *Iron Age* warned its readers, "Some stoker man perhaps will tell

you that his device will do it, but be cautious; sometimes he is fairly successful, but often his machine is very troublesome" (90 [1912]: 19).

28. "Report on the Operations of the Coal-Testing Plant of the United States Geological Survey at the Louisiana Purchase Exposition, St. Louis, MO. 1904," U.S.G.S. *Professional Paper No. 48* (1906), 23–26. Dwight Randall summarized the goals of the U.S.G.S. tests in the introduction to "The Burning of Coal without Smoke in Boiler Plants," U.S.G.S. *Bulletin No. 334* (1908), 5.

29. U.S.G.S. *Professional Paper No. 48* (1906), 29–30.

30. U.S.G.S. *Bulletin No. 325* (1907) and *Bulletin No. 373* (1909), passim.

31. *American Review of Reviews* 39 (1909): 192–95.

32. *Industrial World* 48, pt. 1 (1914): 368, and 44, pt. 2 (1910): 1540–41; Herbert Wilson, "The Cure for the Smoke Evil," *American City* 4 (1911): 263–67. For a brief summary of the establishment of the Bureau of Mines and the evolution of government fuel testing see Guy Elliot Mitchell, "The New Bureau of Mines," *The World To-Day* 19 (1910): 1150–55. Although the bureau assumed the fuel testing duties of the Geological Survey, its primary goal upon its creation was saving the lives of miners. The bureau also offered non-engineering advice to municipalities interested in improving their antismoke departments. In 1912, the bureau published a sample ordinance which emphasized the need for professional engineers to run city smoke departments and the importance of regulating new boiler construction. See Samuel Flagg, "Smoke Abatement and City Smoke Ordinances," United States Bureau of Mines *Bulletin No. 49* (1912), 29–35.

33. Lester Breckenridge, "How to Burn Illinois Coal without Smoke," University of Illinois Engineering Experiment Station *Bulletin No. 15* (1907); "Check on Smoke Evil," Chicago *Record-Herald*, 4 April 1904; *Municipal Engineering* 35 (1908): 127.

34. Breckenridge,"How to Burn Illinois Coal without Smoke," 2. Breckenridge's conclusions concerning the possibility of smokeless combustion of soft coal made news in popular as well as technical publications. *Outlook* reported that the Illinois experiments revealed "clearly and conclusively" that soft coal could be "burned without making objectionable smoke" (*Outlook* 90 [1908]: 54). See also, *The World To-Day* 19 (1910): 1121. At the 1910 meeting of the American Society of Mechanical Engineers, D. T. Randall emphasized the possibility and practicality of smoke abatement. Relying on his own research experience and mounting evidence from other sources, including Breckenridge's work, Randall claimed, "There has been considerable progress made in the design of furnaces during the past five years and there are many plants now operating in which bituminous coal is burned efficiently and without objectionable smoke" (*Transactions of the American Society of Mechanical Engineers* 32 [1910]: 1138).

35. Ernest L. Ohle, "Smoke Abatement—A Report on Recent Investigations Made at Washington University," *Journal of the Association of Engineering Societies* 55 (1915): 139–48. By 1915, Fernald had moved from Washington University to the University of Pennsylvania. He had also taught engineering at Case Western in Cleveland. See

Fernald's "The Smoke Nuisance," *University of Pennsylvania Free Public Lecture Course, 1913–14* (1915), 165–90. Ernest Ohle and Leroy McMaster, "Soot-Fall Studies in Saint Louis," *Washington University Studies* 5, pt. 1 (1917): 3–8.

36. The investigation did not publicize the identity of the donor of the grant. Pittsburgh historian Roy Lubove identified the donor as Robert Duncan Kennedy, who served as a director of the institute, but one of the essays contained in the investigation identified Richard B. Mellon as the source of the grant. The Mellon family was closely associated with the Institute for Industrial Research, which soon bore the family's name. Two Mellon men, Andrew W. and Richard B., joined the committee of the trustees by the 1920s. See Roy Lubove, *Twentieth-Century Pittsburgh* (New York: John Wiley, 1969), 48; Oskar Klotz and William Charles White, eds., "Papers on the Influence of Smoke on Health," *Mellon Institute Smoke Investigation Bulletin No. 9* (1914), 164.

37. A tenth volume appeared in 1922, entitled "Recent Progress in Smoke Abatement and Fuel Technology in England," but this volume lay outside the original investigation's purview.

38. "Outline of the Smoke Investigation," *Mellon Institute Smoke Investigations Bulletin No. 1* (1912), 1–2. O'Connor did not limit his antismoke activities to those related to his work at the Mellon Institute; he also served as the secretary of Pittsburgh's Smoke and Dust Abatement League during that organization's most active years.

39. W. H. Snider to Benner, 24 July 1913; O'Connor to Snider, July 1913; George H. Smith Steel Casting Company to Pittsburgh Investigation, 29 July 1913, all in Smoke Investigation Activities MSS, series 1, folder 2, University of Pittsburgh.

40. John O'Connor, "The Smoke Investigation of the University of Pittsburgh," *Industrial World* 47, pt. 1 (1913): 132; A. A. Straub, "Some Engineering Phases of Pittsburgh's Smoke Problem," *Mellon Institute Smoke Investigation Bulletin No. 8* (1914), 10. The institute sent copies of its bulletin around the nation, and around the world. In February of 1915, O'Connor sent six copies of each bulletin to England's Coal Smoke Abatement Society. See correspondence between Lawrence Chubb and O'Connor, January and February 1915, Smoke MSS, series 1, folder 2, University of Pittsburgh.

41. Washington, Illinois, and Pittsburgh were not the only universities to conduct significant smoke abatement research. Smoke control was an important research field for many engineering programs. The University of Tennessee, for example, conducted extensive tests on steam jets. See *Engineering Magazine* 40 (1910): 406–12.

42. The Civic League of St. Louis, "The Smoke Nuisance," 1906, 25–26.

43. "New Smoke Inspector in St. Louis," *Industrial World* 45, pt. 2 (1911): 925; William A. Hoffman, "The Problem of Smoke Abatement," *Journal of the Association of Engineering Societies* 50 (1913): 250.

44. Chicago *Record-Herald*, 6, 26 September 1907. Bement had gained considerable expertise in smoke abatement engineering. See, for example, his "The Suppression of Industrial Smoke with Particular Reference to Steam Boilers," *Journal of the Western Society of Engineers* 11 (1906): 693–752. In 1912, a similar scenario took place in Cincinnati when that city replaced the highly effective Matthew Nelson with Arthur Hall,

an experienced practical engineer and the highest scorer on a civil service exam. See *Commercial Tribune*, 29 June, 3 July 1912. Also in 1912, at the height of the efficiency craze, Cleveland Mayor Newton Baker requested that his new appointee to chief smoke inspector place the department on an engineering basis. The man he chose for the job, E. P. Roberts, was eminently qualified for the task, possessing many years of experience as a mechanical engineer. Roberts had also served as the chairman of the Chamber of Commerce's influential smoke abatement committee. See "A 'New Deal' for Cleveland on Smoke," *Industrial World* 47, pt. 1 (1913): 134.

Pittsburgh, too, witnessed controversy over the training of its smoke inspector. In 1914, Inspector Searle withstood an attempt to pass an ordinance which required that a college-trained engineer hold his position. Searle gained the support of many practical engineers, including the National Association of Stationary Engineers, who argued that the requirement would unjustly disqualify many exceptional candidates with extensive smoke abatement backgrounds. This battle, then, was not between lay reformers and experts but between two types of experts. See *Industrial World* 48, pt. 1 (1914): 35, 125; *Power* 39 (1914): 201.

45. See for example Raymond C. Benner, "Methods and Means of Smoke Abatement," *American City* 9 (1913): 230–32; J. M. Searle, "Smoke Prevention: The Problem of Cities," *Industrial World* 45, pt. 1 (1911): 858; and "The Gloom of Useless Smoke," *World's Work* 17 (1908): 10865–66, which despite its title concluded that "Chicago is doing well."

46. Osborn Monnett, "New Methods of Approaching the Smoke Problem," *National Engineer* 16 (1912): 720. For a brief discussion of Monnett's career see "Chicago's New Smoke Inspector," *Power* 34 (1911): 230. Monnett served as an editor for *Power*. Pennsylvania Railroad solicitors reported several meetings with Monnett, with whom they developed a friendly relationship. As a sign of good faith Monnett accepted court costs of $46.50 in return for the dismissal of cases involving forty-eight violations of the law. The solicitors concluded, "We feel very gratified to report so satisfactory an outcome of the smoke situation, which was becoming a most serious matter until the appointment of Mr. Monnett as Chief Smoke Inspector" (Loesch, Scofield, and Loesch to W. H. Scriven, 2 August 1911, PA RR MSS, box 1271, folder 16, Hagley Library).

47. Syracuse Chamber of Commerce, "Report upon Smoke Abatement," (1907), 5–12; Emil Pfleiderer to John O'Connor, 21 April, 2 May 1914, Smoke MSS, series 4, folder 7, University of Pittsburgh.

48. Chicago *Record-Herald*, 28, 30 June 1906.

49. For a summary of all of the presentations at the Syracuse meeting see *Power* 31 (1909): 116–18.

50. "Proceedings of the 10th Annual Convention of the International Association for the Prevention of Smoke" (Cincinnati, 1915), 15–17.

51. In 1918 Newark became the first city to hold the convention twice, after eleven other cities had held the honor. The selection of cities seems most related to the par-

ticipation level of smoke inspectors in the association, and was a means of rewarding active inspectors, such as Charles Poethke of Milwaukee, who hosted the second convention in 1907. Oddly, Chicago did not host a convention before the end of World War I, even though its inspectors were very active in the association.

52. *Industrial World* 47, pt. 2 (1913): 1092.

53. International Association for the Prevention of Smoke, "Third Annual Convention" (Milwaukee, 1908); Smoke Prevention Association, "Proceedings of the Tenth Annual Convention" (Cincinnati, 1915); "Proceedings of the Eleventh Annual Convention" (St. Louis, 1916); "Thirty-Third Annual Convention" (Milwaukee, 1939). Other conventions are discussed at length in *Industrial World* and *Smoke*, the association's official bulletin.

54. Milwaukee *Sentinel*, 26 June 1907. Perhaps fearing the consequences of the exclusive nature of the Smoke Prevention Association, John O'Connor wrote to William Hoffman, the chief smoke inspector of St. Louis, in an effort to ensure broad participation at the convention to be held in that city in 1916. O'Connor asked that the association make a special effort to attract representatives from active civic organizations, specifically the smoke abatement leagues of St. Louis, Louisville, Cincinnati, and his own Smoke and Dust Abatement League (O'Connor to Hoffman, 22 July 1916, Smoke MSS, series 4, folder 6, University of Pittsburgh).

55. "Proceedings of the Eleventh Annual Convention" (St. Louis, 1916), passim.

56. McKnight's Women's Health Protective Association merged with the Civic Club of Allegheny County when that organization formed in 1895. McKnight also helped form the Twentieth Century Club, an organization of prominent women, which, along with the Civic Club, joined the Smoke and Dust Abatement League in 1912 in a concentrated effort to improve the city's air.

57. Pittsburgh *Post*, 10 November 1906. The engineering argument against smoke gained so much currency by the early 1910s that a Baltimore women's organization used the call for efficiency as the title of a smoke abatement pamphlet: "Black Smoke Means Waste of Fuel and Loss of Boiler Efficiency: History of the Work of the Smoke Committee of the Women's League" (Baltimore, no date), Smoke MSS, series 4, folder 9, University of Pittsburgh.

58. New York *Times*, 15 August 1913.

six Smoke Means Waste

1. Tapping into the salience of conservation, one writer even referred to smoke abatement as the "conservation of the purity of the air." For the author, clean air had joined the list of natural resources, plundered by years of greed and left in seriously short supply. For most observers, however, the resource wasted in the creation of smoke was not pure air, but coal. See Alexander G. McAdie, "Conservation of the Purity of the Air—Prevention of Smoke," *Monthly Weather Review* 38 (1910): 1423.

2. Oskar Klotz and William Charles White, eds., "Papers on the Influence of Smoke on Health," *Mellon Institute Smoke Investigation Bulletin No. 9* (1914), 164–73.

3. See Michael Teller, *The Tuberculosis Movement* (New York: Greenwood Press, 1988), and Barbara Bates, *Bargaining for Life* (Philadelphia: University of Pennsylvania Press, 1992). See also *Journal of the Outdoor Life*, an activist magazine dedicated to the treatment of tuberculosis, which through the most active years of the campaign, 1906–1918, dedicated very little space to the smoke issue.

4. Historian Nancy Tombs has argued that "the germ theory did not vitiate the connection between a clean, moral life and safety from disease." Tombs's work focuses primarily on the 1870s and 1880s, when germ theory still coexisted with miasmatic theory in public conceptions of disease. But by the second decade of the 1900s, these Victorian connections had deteriorated. See "The Private Side of Public Health: Sanitary Science, Domestic Hygiene, and the Germ Theory, 1870–1900," *Bulletin of the History of Medicine* 64 (1990): 529–30.

5. In a letter to the company solicitors in Chicago, Pennsylvania's General Agent and Superintendent W. H. Scriven wrote, "It is evident that this crusade for the prevention of smoke is being waged against the railroads with the idea of persecuting them to such an extent as to secure the electrification of the roads" (Scriven to Loesch, Scofield, and Loesch, 26 February 1911, Pennsylvania Railroad MSS, box 1271, folder 16, Hagley Library).

6. Paul Bird, "Locomotive Smoke in Chicago," *Railway Age Gazette* 50 (1911): 321. The position of the Illinois Central's line through Grant Park became a major issue for some Chicagoans. One Chicago *Record-Herald* editorial read: "Many portions of the city are befouled by locomotive smoke, but the spot of all spots which needs protection is the lake front of the business district. The public park lands there are made useless by smoke, cinders and soot. One cannot even walk the length of the park without getting more injury from smoke than he gets benefit from exercise and air" (26 January 1906).

7. Chicago *Record-Herald*, 4, 26 January 1906.

8. Cloyd Marshall, "Electric Traction," *Railway and Engineering Review* 38 (1897): 745–46. As Marshall spoke only one steam railroad had undertaken an electrification project: Baltimore and Ohio had opened an electrified line through a tunnel in Baltimore. Most early electrification projects involved streetcars and interurban railways, not the larger steam railroad lines. See Michael Bezilla, *Electric Traction on the Pennsylvania Railroad* (University Park: Pennsylvania State University Press, 1980), 3–6.

9. *Railway and Engineering Review* 37 (1897): 189. For background into electric traction see Carl W. Condit, *The Port of New York: A History of the Rail and Terminal System from the Beginnings to Pennsylvania Station* (Chicago: University of Chicago Press, 1980), 176–238.

10. For a positive assessment of electric traction technology as it existed in 1905, see Condit, *The Port of New York*, 229–38. Many of the strongest proponents of electrification had an economic interest in the new technology, including George Westing-

house, who stood to gain considerable business if railroads adopted his electrical system. Westinghouse, who championed alternating current in competition with General Electric's direct current, stressed the major advantage of his system over others: alternating current was more cost effective over long rail lines. Westinghouse envisioned the complete electrification of railroads, including cross-country freight lines. See George Westinghouse, "The Electrification of Steam Railways," *Railway Electrical Engineer* 2 (1910): 52–55; also General Electric's George W. Cravens, "Electrification of Steam Railroads," *Railway Electrical Engineer* 1 (1909): 164–65.

11. New York *Times*, 9, 10 January 1902; *Railroad Gazette* 34 (1902): 18.

12. New York *Times*, 10 January 1902. Just the previous year discomfort to passengers riding through the Fourth Avenue tunnel led to a grand jury investigation. The grand jury concluded, among other things, "[t]hat the motive power in the tunnel [should] be changed to avoid the use of coal." Shortly thereafter, the *Railroad Gazette* editorialized that "the art of electric traction . . . appears to have reached a state where radical improvement is not only possible but practicable" (*Railroad Gazette* 33 [1901]: 565, 576).

13. New York *Times*, 10 January, 11 February 1902. The coroner's investigation, completed within a month of the accident, exonerated the engineer, blaming instead the New York Central for allowing smoky and steamy conditions to persist in the tunnel. See *Railroad Gazette* 34 (1902): 78.

14. Bezilla, *Electric Traction on the Pennsylvania Railroad*, 18–25; Henry Obermeyer, *Stop That Smoke!* (New York: Harper, 1933), 59. See also Kurt Schlichting, "Grand Central Terminal and the City Beautiful in New York," *Journal of Urban History* 22 (1996): 332–49. Schlichting places the Grand Central improvements within the City Beautiful movement, discussing particularly the architecture of the many new buildings, including the terminal itself, which accompanied electrification. Unfortunately, Schlichting fails to include a discussion concerning smoke, steam, and noise abatement, all of which would strengthen his argument concerning the relationship between the project and City Beautiful.

15. Bezilla, *Electric Traction on the Pennsylvania Railroad*, 56–73. Philadelphia's smoke inspector, John M. Lukens, claimed that pressure from the city to abate smoke had aided in Pennsylvania Railroad's decision to electrify their lines. In the relatively smoke-free city of Philadelphia the smoke-belching locomotives of the city's most important railroad made the company an obvious target of antismoke ire and action, and, concluded Lukens, "it is likely that the growing sentiment against the smoke nuisance has played an important part in the decision of the Pennsylvania Railroad to announce the coming electrification of the Germantown and Chestnut Hill Division, and of the Main Line as far out as Paoli." Given the company's resistance to electrification in Chicago and other smoky cities, however, direct economic considerations undoubtedly determined the electrification decision. See *Third Annual Message of Rudolph Blankenburg, Mayor of Philadelphia* (1913), 532–33.

16. A. W. Gibbs, "The Smoke Nuisance in Cities," *Railroad Age Gazette* 46 (1909):

412–415. Gibbs's paper was a response to an American Civic Federation request for information on the smoke problem vis-à-vis railroads.

17. David F. Crawford, "The Abatement of Locomotive Smoke," *Industrial World* 47, pt. 2 (1913): 1095–100. The cost of electrification was indeed "tremendous" in New York City. The Pennsylvania Railroad estimated the cost of its entire electrification project in New York, including the building of the tunnel, at $159 million (American Society of Civil Engineers, *Transactions* 68 [1910], 9). The Pennsylvania Railroad made no attempt to electrify in Pittsburgh, where the company hired eight smoke inspectors to watch its locomotives for smoke. These company smoke inspectors could instruct firemen on smokeless operation and report persistent violations to company officials, who in turn reprimanded or even suspended ineffective firemen (J. M. Searle, "Report of Division of Smoke Inspection, Pittsburgh, PA.," *Industrial World* 44, pt. 2 [1910], 1151–52).

18. Condit, *The Port of New York*, 176–238.

19. Rea estimated the total cost of improvements in D.C., without electrification, at $20 million (Senate Committee on the District of Columbia, "Statements before the Committee on H.R. 9329," 1907, 20–25; House, "Additional Terminal Facilities at the Union Station," 59th Cong., 1st sess., 1906, H. Doc. 2563, 1–2).

20. George Gibbs to W. W. Atterbury, 11 January 1906, PA RR MSS, box 153, folder 31, Hagley Library. Gibbs did seem to assume that all lines would eventually be electrified. The only question involved timing.

21. Rea to McCrea, 30 January 1907, PA RR MSS, box 153, folder 31, Hagley Library. Rea did order the use of smokeless fuel and additional care from enginemen and firemen (Rea to Atterbury, 26 January 1907).

22. Washington *Evening Star*, 20 February, 1 March 1907, as found in PA RR MSS, box 153, folder 32, Hagley Library.

23. See correspondence among Samuel Rea, George Stevens, W. H. White, February and March 1907; quote from Rea to Stevens, 30 March 1907, PA RR MSS, box 153, folder 32, Hagley Library.

24. E. F. Brooks to W. W. Atterbury, 29 November 1907, PA RR MSS, box 153, folder 32, Hagley Library. Electrification between New York and Washington came in pieces, and the first regularly scheduled electric train to run from New York to Washington arrived in February 1935. See "1,343 Miles of Electrified Track," *Fortune* 13 (June 1936): 94–98, 152.

25. Chicago *Record-Herald*, 31 March, 17 April 1908.

26. Chicago *Record-Herald*, 12 September 1908. Historian Harold Platt also offers an interpretation of Chicago's electrification movement of 1908–15. Although he articulates fine arguments on a complex subject, Platt's lack of national context leads him to a dubious conclusion. Platt argues that Sergel and the women of the Anti-Smoke League offered a novel definition of the smoke problem as they shifted the public discourse on smoke from largely technical terms to largely health and aesthetic terms. As discussed above, women in many American cities, including Cincinnati, St.

Louis, and Pittsburgh, had been at the forefront of antismoke activism and had established the dominant definition of the smoke problem as early as the 1890s. In 1908, the women of Chicago offered neither "originality" nor a redefinition of the "terms of the policy debate," as Platt concludes. Although Chicago was among the first cities to move to a professional (engineering) staff in its smoke inspection department—through the department's re-creation in 1903 under the Department for the Inspection of Steam Boilers and Steam Plants—that city had also defined the smoke problem largely in health and aesthetic terms. In 1893, a movement to abate smoke developed around the aesthetic preparation of the city for the World's Fair.

Platt also overemphasizes the role of germ theory in his discussion of the 1908 movement in Chicago. Germ theory was far from new in 1908, as Platt notes, and it played no significant role in the women's decision to attack locomotive smoke as a health threat and had very little to do with government action and business reaction. See Platt, "Invisible Gases: Smoke, Gender, and the Redefinition of Environmental Policy in Chicago, 1900–1920," *Planning Perspectives* 10 (1995): 67–97.

27. Chicago *Record-Herald*, 17, 18 September 1908. The city also undertook a new feasibility study of electrification and prepared to issue its report, which contradicted railroad claims that electrification would not decrease operational costs or allow for increased traffic.

28. Chicago *Record-Herald*, 19 September 1908.

29. Chicago *Record-Herald*, 22, 24, 25, 26 September and 6, 14 October 1908. The *Record-Herald* had quoted Fritch as saying, "No railroad would be justified in spending many millions of dollars, unless it can be demonstrated to be good business to do so."

30. Chicago *Record-Herald*, 6, 20 October 1908. A *Record-Herald* editorial referred to the work of Dr. John Wainwright which appeared in New York's *Medical Record* 74 (1908): 791, during the electrification debate in Chicago. Wainwright gave support to the argument that smoke adversely affected health, allowing the *Record-Herald* to write: "It impairs and destroys the tissues of the nose, throat, eyes, lungs and air passages; it aggravates the discomfort of those suffering from heart disease; it intensifies nervous disorders; it is a peril to the aged; it lowers the tone of general health, and reduces still further an already lowered resistance to disease" (19 November 1908). Although the focus of the Anti-Smoke League remained the Illinois Central, obviously other railroads created their own smoke nuisances in other neighborhoods. Sergel helped inspire the members of the Englewood Woman's Club to initiate their own campaign against the Rock Island Railroad, using means similar to those of the Anti-Smoke League (*Record-Herald*, 24 November 1908).

31. Chicago *Record-Herald*, 4, 19 December 1908. If the railroads often failed to cooperate in the city's effort to control smoke emissions, they did cooperate with each other. By 1915, J. H. Lewis, a smoke inspector employed by a Chicago railroad, could report to the International Association for the Prevention of Smoke that railroad smoke inspectors in his city had formed an association. Inspectors from thirty-

three railroads in the city shared reports and observations, so that a Rock Island inspector, for example, would notify a Northwestern inspector if he saw a Northwestern locomotive in violation of the ordinance. In this way, the railroads had created a network of track observers in an effort to control smoke and avoid public interference. See International Association for the Prevention of Smoke, "Proceedings" (1915), 39.

32. *Locomotive Firemen's Magazine* 32 (1902): 156–159; Chicago *Record-Herald,* 29 June 1911. The railroads disciplined a significant number of firemen and engineers for smoke violations. In early November 1909, the *Record-Herald* reported that fifty men complained to their unions about such discipline. The Brotherhood of Engineers responded by offering newspapers photographs of railroad yards which were relatively smokeless (2 November 1909).

33. Chicago *Record-Herald,* 6, 7 January 1909.

34. Chicago *Record-Herald,* 8 July and 21, 25 October 1909.

35. "Proceedings of the Local Transportation Committee, City Council of Chicago, November 17, 1909," 5–7, as found in PA RR MSS, box 1271, folder 15, Hagley Library. This record appears to be the work of a stenographer hired by the Pennsylvania Railroad, and it differs slightly from quotations made in the *Record-Herald,* in which the "sacrifice" quote appeared (18 November 1909).

36. At least one opponent of strict smoke abatement went straight to the Anti-Smoke League. C. M. Moderwell, president of the United Coal Mining Company, met with Sergel and other women, hoping to convince them that coal could be burned without smoke. Moderwell apparently had little effect on the women's impressions of soft coal. See Chicago *Record-Herald,* 29 October and 4, 18 November 1909; "Proceedings . . . November 17, 1909," 13, 33–36. Here the quotation from Kelley is different from that which appears in the *Record-Herald.* It reads: "I believe that this is a woman's law, an anti-smoke nuisance composed mostly of women."

37. "Proceedings . . . November 17, 1909," 59–61.

38. Chicago *Record-Herald,* 29 November 1909.

39. Chicago *Record-Herald,* 7 November 1909. Rend had also exerted influence through his participation in an organization with a vested interest in the smoke issue, the Illinois Coal Operators Association. He served on the association's Coal Stoking and Anti-Smoke Committee, which in 1908 published a pamphlet concerning the smokeless combustion of Illinois and Indiana coals and separate instructional posters for firemen, on protected cardboard so that they might hang in boiler rooms. The coal operators' activism reflected their understanding of the importance of achieving smokelessness for their product and, perhaps, their understanding of the need to show concern for the smoke issue. Serving with Rend on the antismoke committee was Alburto Bement, a member of the city's smoke abatement advisory board, which supplied Smoke Inspector Bird with technical guidance. Bement's participation in the project no doubt greatly aided in attracting the support of the Department of Smoke Inspection and the Western Society of Engineers, both of which assisted in the publi-

cation of the pamphlet. See Joint Committee on Smoke Suppression, "The Use of Illinois and Indiana Coal without Smoke" (1908).

40. D. C. Moon to A. F. Banks et al., 28 January 1910; "Report of the Meeting of Railroad Officials Held at Chicago, February 27th, 1911," box 1271, folder 15; W. D. Cantillon to G. L. Peck, 11 April 1911, box 1271, folder 16, PA RR MSS, Hagley Library.

41. Chicago *Record-Herald*, 23 December 1909 and 16 February, 26 March 1910. Several Chicago lines also invested in a new switching facility outside the city proper. Aside from placing the new switch yard beyond the reach of any future city council legislation requiring electrification, the new facility would also reduce the amount of smoke produced by locomotives in the city, particularly given the notorious reputation of switching cars as heavy smokers. See Chicago *Record-Herald*, 1 December 1908 and 3, 7 April 1911.

42. Rea to Turner, 20 August 1912, PA RR MSS, box 153, folder 33, Hagley Library. By the time Rea wrote this letter, General Manager G. L. Peck had already requested and received the company's records concerning the Washington situation. Peck hoped that they might contain suggestions on how to handle the Chicago movement (W. H. Myers to G. L. Peck, 10 December 1909, box 1271, folder 15).

43. W. A. Garrett to Cantillon et al., 18 September 1912, PA RR MSS, box 1271, folder 17, Hagley Library. Interestingly, the subcommittee unanimously agreed that it was not advisable to form a joint smoke inspection bureau less than a week later, but the association apparently ignored the recommendation.

44. Joint Smoke Inspection Bureau of the Railroads Operating in Chicago, "Instructions for Making Reports," 17 December 1912, PA RR MSS, box 1271, folder 17; Chicago General Managers' Association, 28 May 1914, PA RR MSS, box 1272, folder 3, Hagley Library.

45. Chicago Association of Commerce, *Smoke Abatement and Electrification of Railway Terminals in Chicago* (1915), 19–23.

46. Chicago Association of Commerce, *Smoke Abatement and Electrification*, 19–23; W. F. M. Goss, "Smoke as a Source of Atmospheric Pollution," *Journal of the Franklin Institute* 181 (1916): 306; Chicago *Tribune*, 20 May 1913.

47. Chicago *Record-Herald*, 2 December 1911 and 30 March 1912. Certainly the women had to defer to engineers and economists on many aspects of the electrification debate. But on one crucial issue, the timing of electrification, the women were as expert as any other group. Although the women of the Anti-Smoke League had not offered a detailed smoke solution, they had determined that electrification would be the only permanent and complete answer to railroad smoke. Thus, for the women the only pertinent question became when the city would force electrification. For the engineers and economists of the association study, and by late 1911, the mayor, several other questions needed to be answered, particularly concerning the economic desirability of electrification.

48. Chicago *Tribune*, 12, 13 May 1913. Eight railroads sent representatives to the committee: Burlington, Northwestern, St. Paul, Pennsylvania, Illinois Central, New

York Central, Santa Fe, and the Rock Island. In early 1913 the railroads were spending about $15,000 per month on the C.A.C. study. The costs were divided among the roads using a formula of track mileage and number of passenger cars entering the city. Under this arrangement the Chicago and Northwestern paid the most, averaging just over $2,000 per month for four years. The Illinois Central paid about $1,800 per month. Railroads made their payments through the General Managers' Association. See W. S. Tinsman to B. McKeen, 18 April 1913, PA RR MSS, box 1272, folder 1, Hagley Library.

49. Chicago *Tribune*, 20 May 1913.

50. Chicago *Tribune*, 3 June 1913.

51. Chicago *Tribune*, 3 November 1915; Chicago Association of Commerce (1915), 1052. The report also included a table identifying the total amount and type of coal consumed by railroads in Chicago in 1912. The figures reveal that even after years of promising to switch to smokeless fuels, such as coke or anthracite, or even the less smoky Pocahontas coal from West Virginia, the city's railroads still relied heavily on the dirty bituminous coals of Illinois and Indiana. Over 72 percent of the coal consumed by locomotives in the city was bituminous (*Journal of the Franklin Institute* 181 [1916]: 318).

52. One *Record-Herald* cartoon even depicted women sitting atop office buildings knitting and watching stacks for violations. Although unlikely to sit on each building, women did systematically watch stacks around Chicago and other cities (Chicago *Record-Herald*, 19 April 1909). See activities of the Cincinnati Woman's Club outlined in chapter 3 and figure 3 (p. 53).

53. Chicago Association of Commerce (1915), 173; Chicago *Record-Herald*, 16 February 1911. The report's arguments about the contribution of locomotive emissions to the total smoke cloud of Chicago sound remarkably similar to more recent arguments about contributions of various kinds of garbage to landfills. By complicating the issue, by including atmospheric dust and all sources of smoke, the study minimized the impact that electrification would have on total air quality.

54. The association relied on published material for this section, particularly the Mellon investigation bulletins, which Goss requested from O'Connor as they became available in 1913 and 1914. See correspondence between W. F. Goss and John O'Connor, 23 July 1913 to 19 May 1914, Smoke Investigation Activities MSS, series 1, folder 21, University of Pittsburgh Library.

55. The amount of information requested by the C.A.C. surprised even Pennsylvania General Manager G. L. Peck, who wrote after receiving one questionnaire, "This seems to be pretty elaborate information these people want and I do not know what they want it for, but I suppose you have it in convenient shape" (Peck to W. H. Scriven, 26 March 1912, PA RR MSS, box 1271, folder 17, Hagley Library). See box 1272, folder 1 for railroad questionnaires from Horace G. Burt of the C.A.C. study.

56. Chicago Association of Commerce (1915), 1051–52.

57. Goss, "Smoke as a Source of Atmospheric Pollution," *Journal of the Franklin In-*

stitute 181 (1916): 305–38. Goss also spoke before the Engineers' Society of Western Pennsylvania in 1915, and his speech was reprinted in *Steel and Iron* 49, pt. 1 (1915): 224–26. Goss was a respected engineer with a background in locomotive smoke. He had also served on the first Association of Commerce committee, and had published a bulletin with the Bureau of Mines concerning fuel tests on locomotives (W. F. M. Goss, "Comparative Tests of Run-of-Mine and Briquetted Coal on Locomotives," Bureau of Mines *Bulletin No. 363* [1908]). For a brief biography of Goss see *Municipal Engineering* 45 (1913): 151.

58. "Where Smoke Really Comes From: The Truth about the Smoke Nuisance," reprinted by the publicity department of the Pennsylvania Railroad System, from *Railway and Locomotive Engineering* (January 1916), 3–4. Even accepting the association's assertion that railroads contributed only 22 percent of the city's smoke, the complete abolition of this percentage would have meant a remarkable difference in the clarity of the air, particularly along the lake front and near railroad terminals. The Pennsylvania's pamphlet drew the attention of the Cincinnati *Times-Star*, which ran an editorial denouncing the railroad. "The very issuance of such a pamphlet, which is nothing but a clumsy avoidance of the smoke question, brings under suspicion the sincerity of the Pennsylvania railroad in the abatement of the smoke nuisance," it read in part. The editorial was so strongly worded that General Manager B. McKeen asked David Crawford to send representatives from Motive Power to discuss the matter with editor Charles Taft. Taft assured the Pennsylvania men that he did not write the editorial. See McKeen to Crawford, 7 April and Crawford to McKeen, 29 April 1916, PA RR MSS, box 1279, folder 1, Hagley Library; Cincinnati *Times-Star*, 15 March 1916.

59. "A Remarkable Exhibit of Railway Apparatus," *Electric Journal* 12 (1915): 477–80. The highly praised Panama-Pacific Exposition showcased the miraculous rebirth of a city all but destroyed only nine years earlier, when an earthquake touched off a massive fire that consumed the city.

60. Chicago *Tribune*, 3 December, 4 November 1915; John F. Stover, *The History of the Illinois Railroad* (New York: Macmillan, 1975), 298–99.

61. For a thorough description of Birmingham see Marjorie Longenecker White, *The Birmingham District: An Industrial History and Guide* (Birmingham: Birmingham Historical Society, 1981), esp. 46–64. See also Carl V. Harris, *Political Power in Birmingham, 1871–1921* (Knoxville: University of Tennessee Press, 1977), 12–38.

62. Quoted in Harris, *Political Power in Birmingham*, 229.

63. W. David Lewis, *Sloss Furnaces and the Rise of the Birmingham District* (Tuscaloosa: University of Alabama Press, 1994), 324–25.

64. Birmingham *Age-Herald*, 17 January 1913.

65. The *Age-Herald* consistently used "pay roll" rather than "payroll" during the smoke abatement debate.

66. Birmingham *Age-Herald*, 17, 28 January 1913.

67. "Soft Coal Bound to Produce Smoke," Birmingham *Age-Herald*, 28 January 1913.

68. "Smoke Ordinance Impracticable," Birmingham *Age-Herald*, 28 January 1913; "Smoke Ordinance Is Last Straw; Small Plants Prepare to Leave," *Age-Herald*, 30 January 1913; "More Small Industries Needed," *Age-Herald*, 4 February 1913.

69. Birmingham *Labor-Advocate*, 31 January 1913.

70. Joel A. Tarr, "Searching for a 'Sink' for an Industrial Waste: Iron-Making Fuels and the Environment," *Environmental History Review* 18 (1994): 9–34.

71. White, *The Birmingham District*, 63; Birmingham *Labor-Advocate*, 14, 21 February 1913.

72. Birmingham *Age-Herald*, 1 February 1913.

73. Birmingham *Age-Herald*, 30 January, 4 February 1913.

74. "Can See No Difference in Atmosphere of City," Birmingham *Age-Herald*, 4 February 1913; "Smoke Not as Bad as Dust, Say Well Known Citizens," *Age-Herald*, 4 February 1913.

75. Birmingham *Age-Herald*, 8, 22, 26 February 1913.

76. Birmingham *Age-Herald*, 1 March 1913; Lewis, *Sloss Furnaces and the Rise of the Birmingham District*, 325; Harris, *Political Power in Birmingham*, 230. Of course Birmingham was not the only city that failed to create an effective means of regulating smoke emissions. In other young industrial cities, particularly where one or two major employers dominated payrolls and politics, significant regulation of industrial pollution gained little public support. In Gary, Indiana, for example, United States Steel so completely dominated the city that municipal officials did not attempt the regulation of its smoke emissions, even though the city had passed an antismoke ordinance in 1910. In Butte, Montana, as well, smoke abatement efforts had little effect. As early as the 1890s, local women and a physician had lobbied for municipal action against the intense air pollution from the smelters in the city. Although the activism did lead to some restrictions on the most polluting ore roasting process, a fear that an active regulation of smelters would lead to a loss of jobs prevented real reform of the industry. Indeed, not until the removal of several of the city's smelters did Butte see real relief from the poisonous smelter emissions. See Andrew Hurley, *Environmental Inequalities* (Chapel Hill: University of North Carolina Press, 1995), 38–39; Donald MacMillan, "A History of the Struggle to Abate Air Pollution from Copper Smelters" (Ph.D. diss., University of Montana, 1973), 13–43.

seven War Meant Smoke

1. Pittsburgh *Gazette Times*, 23 October 1916; H. M. Wilson to William Todd, 6 July 1912, Smoke Investigation Activities MSS, series 3, folder 1, University of Pittsburgh Library.

2. Pittsburgh *Gazette Times*, 24 October 1916; *Power* 44 (1916): 671. Monnett had a busy schedule during the week. He also spoke before the National Association of Stationary Engineers and a joint meeting of the Rotary and Civic clubs.

3. See International Association for the Prevention of Smoke, "Proceedings" (1915

and 1916). Certainly Hood's expertise went unquestioned, given his work at the Bureau of Mines.

4. Pittsburgh *Gazette Times*, 25 October 1916.

5. Pittsburgh *Gazette Times*, 24 October 1916.

6. The shift in antismoke rhetoric from health-beauty-cleanliness toward efficiency did not require a refutation of old ideas. Rather, the increasing importance of efficiency issues signaled the shifting focus from public environmental concerns toward private conservation interests.

7. Lonergan's report on the smoke conditions in New York were reprinted in the *Heating and Ventilating Magazine* 14 (October 1917): 29. Daniel Maloney, "In the Interest of Smoke Regulation and Coal Conservation" (1918), 1–2.

8. *Power* 47 (1918): 565; *Survey* 40 (1918): 45. Although the league desired posters to connect smoke abatement to fuel conservation, some posters merged both major themes of smoke abatement, combining arguments concerning aesthetics and conservation. "Fight for Beauty and Democracy—Help Abate Smoke and Conserve Fuel—Make Your City Beautiful and Help Win the War," demanded one poster (Smoke MSS, series 3, folder 20, University of Pittsburgh Library).

9. Pittsburgh Bureau of Smoke Regulation, "Hand Book for 1917," 3.

10. New York *Times*, 13 December 1917. See John G. Clark, *Energy and the Federal Government* (Urbana: University of Illinois Press, 1987), 50–88, for a discussion of the energy crisis and the federal reaction.

11. James Johnson, *The Politics of Soft Coal* (Urbana: University of Illinois Press, 1979), 63; New York *Times*, 15 November 1917 and 2, 9 January 1918.

12. New York *Times*, 13, 21 December 1917 and 18 January 1918; "Finding Coal for Empty Buckets," *Survey* 39 (1918): 449. Pneumonia was not the only smoke-related issue transformed by the fuel crisis. In February 1918, as the nation began to emerge from the nadir of the coal shortage, E. W. Rice, president of General Electric, urged the nation to electrify all railroads as a means of solving the coal shortage and the transportation problem. Electrification, Rice argued, would save the nation one hundred million tons of coal per year, due to increased efficiency. In addition, the more powerful electric locomotives would greatly increase the hauling capacity of individual trains. See "Urges Electric Power for All Railroads," New York *Times*, 16 February 1918.

13. Even as the government prepared to alleviate the coal shortage, Henderson lobbied for broad federal action on the issue, attempting to ensure federal action involving the coal supply would also involve smoke abatement. Henderson argued that smoke regulation should be under federal control because "[i]ts importance is too great to be left within the province of cities, counties or states" ("Smoke Abatement Means Economy," *Power* 46 [1917]: 126–27).

14. New York *Times*, 19 January 1918.

15. At the time of his appointment, Garfield was the president of Williams College. As an old academic acquaintance of Wilson's, he had little experience with the

coal industry. See James Johnson, "The Wilsonians as War Managers: Coal and the 1917–18 Winter Crisis," *Prologue* 9 (1977): 193–208.

16. Johnson, *The Politics of Soft Coal*, 48–70 passim, 70–73. The coal consumption of locomotives was no small issue. In 1917, locomotives consumed 21 percent of the nation's total coal production. See Illinois Engineering Experiment Station, "The Economical Use of Coal in Railway Locomotives," *University of Illinois Bulletin* 16 (September 9 1918): 9.

17. Johnson, *The Politics of Soft Coal*, 67–69; New York *Times*, 18 26 January 1918. For a discussion of the negative reaction to Garfield's factory shutdown see James Johnson, "The Fuel Crisis, Largely Forgotten, of World War I," *Smithsonian* 7 (1976): 64–70. Garfield's industrial shutdown order outraged many members of the Senate. Nebraska's Hitchcock called it a "national disaster," and Mississippi's Vardaman commented, "It strikes me that this order is without justification under the circumstances." The Senate discussed the order at length and then approved a resolution demanding a five-day delay in its implementation so that its effects might be discussed in more detail. See *Congressional Record*, 65th Cong., 2nd sess., 1918, 56, pt. 1, 912–22, 928–36.

18. "Alarming Fuel Situation Is Described," *Iron Age* 101 (1918): 1546–47; James Garfield, *Final Report of the United States Fuel Administrator* (Washington, D.C., 1919), 246–47 250–51. Garfield's report also contains the report of Augustus Cobb, the acting director of the Bureau of Conservation.

19. Garfield, *Final Report*, 248–50.

20. Ibid., 248–55. Breckenridge had moved from the engineering department at the University of Illinois to Yale University. See L. P. Breckenridge, "The Conservation of Fuel in the United States: An Outline for a Proposed Course of Lectures in Higher Educational Institutions" (1918), 7, 68. The Fuel Administration was not the only governmental body active in promoting fuel conservation. As part of its continuing effort to improve the nation's coal consumption efficiency, the Bureau of Mines published a poster addressed to firemen during the war. With an American flag and the capital building in the background, the poster read, "The man with the shovel controls the wise use of coal. The country needs his skill to prevent waste of fuel to keep us warm and to turn the wheels of industry that shall command victory" (Smoke MSS, series 3, folder 21, University of Pittsburgh Library).

21. Pennsylvania geologist Franklin Pratt had expressed concerns about the waste of anthracite in mining and processing as early as 1881, claiming that less than two-thirds of the available coal was ever taken out of the mines, with the remainder left to support mine ceilings. See Pratt, *Second Geological Survey of Pennsylvania* (1881), 39. Gifford Pinchot's concern for the coal supply also involved the waste of coal at the mine. Mining interests, treating coal as if it were inexhaustible, had mined coal inefficiently, Pinchot thought, taking only the most accessible parts of veins and leaving the remainder. See Pinchot, *The Fight for Conservation* (New York: Doubleday, Page, 1910), 6–7.

22. Breckenridge, "The Conservation of Fuel in the United States," 75–76. Although smoke abatement was not a major concern of the Fuel Administration, coal conservation was, and the nation joined in the fuel conservation crusade, often rechanneling the concern once attached to the smoke abatement movement. The trade publication *Heating and Ventilating Magazine*, for example, which had in previous years published a "Smokeless Boiler Edition," replaced it with the "Fuel Conservation Edition," in 1918. Rather than focusing on smokeless combustion, the edition included all manner of information concerning the coal situation (*Heating and Ventilating Magazine* 15 [1918], 44–45).

23. "To Save Coal by Using Daylight," *Literary Digest* 56 (February 16 1918): 14; "The Social Benefits of Daylight Savings," *Survey* 39 (1918): 420; "Why Moving Clocks ahead Will Help Win the War," *Current Opinion* 64 (1918): 366–67; House Committee on Interstate and Foreign Commerce, *Daylight Saving*, 65th Cong., 2nd sess., 1918, H. Rept. 293, 1–12; New York *Times*, 3, 16, 20, 31 March 1918.

24. Frank Haynes, the president of the United Mine Workers, remarked in January 1918, "It is too bad the American people are suffering the hardships of a coal famine when they have an inexhaustible supply of coal almost at their doors" (New York *Times*, 17 January 1918).

25. Carl Harris, *Political Power in Birmingham, 1871–1921* (Knoxville: University of Tennessee Press, 1977), 230; *Power* 47 (1918): 6; 45 (1917): 461; 46 (1917): 330; Smoke Abatement League of Hamilton County, "Report of the Superintendent, 1916" (1917), 7; Walter Pittman, "The Smoke Abatement Campaign in Salt Lake City, 1890–1925," *Locus* 2 (1989): 73.

26. "City Will Have Winter of Smoke," Milwaukee *Sentinel*, 25 November 1918.

27. Philadelphia, *Annual Reports* (1917), 1:300.

28. Philadelphia, *Annual Reports* (1917), 1:299; (1918), 1:283; Osborn Monnett, "Smoke Abatement," Bureau of Mines *Technical Paper 273* (1923), 1.

29. "Pittsburgh Smoke Regulation in 1918," *Power* 49 (1919): 469; Smoke Abatement League of Hamilton County, "Report of the Superintendent, 1916" (1917), 7; John O'Connor to D. W. Kuhn, 22 January 1918, Kuhn to O'Connor, 31 January 1918, Smoke MSS, series 3, folder 10, University of Pittsburgh Library.

30. *Power* 45 (1917): 461. On top of all of the complications brought on by the war, the smoke abatement movement also lost one of its most influential figures and consistent voices, when Pittsburgh's Henderson died in late 1918. See *Power* 49 (1919): 470.

31. Pittsburgh *Post*, 15 July 1917.

32. William G. Christy, "History of the Air Pollution Control Association," *Journal of the Air Pollution Control Association* 10 (April 1960): 128; "Smoke Prevention Association Proceedings, 12th Annual Convention, Columbus, OH, 1917," 70–71.

33. *Power* 48 (1918): 398–400; O'Connor to Hoffman, 22 July 1916, Smoke MSS, series 4, folder 6, University of Pittsburgh Library.

34. *Power* 48 (1918): 398–400. The only presentation from a government official came from Robert Collett, of the Conservation Department of the Fuel Administration.

35. Chicago *Record-Herald*, 25 November 1908.

36. Of course, urbanites had much more evidence connecting impure water and sewage with disease than they did relating air pollution to health concerns, which explains the remarkable success in building public waterworks in the decades before World War I. See Stuart Galishoff, "Triumph and Failure: The American Response to the Urban Water Supply Problem, 1860–1923," in *Pollution and Reform in American Cities, 1870–1930*, ed. Martin Melosi (Austin: University of Texas Press, 1980).

eight Movement toward Success

1. From Milwaukee Legislative Reference Library Collection, Milwaukee *Journal*, 23 November 1928; 10 March 1929; 28 April, 7 September 1932; 23 March 1933.

2. Milwaukee *Journal*, 19 November, 1 December 1935; 15 January 1941.

3. From Women's City Club of Cleveland *Scrapbook*: Cleveland *Plain Dealer*, 19, 20 February and 11, 12 March 1922; Cleveland *Press*, 27, 28 February and 16 March 1922; Cleveland *News*, 2 December 1922, all in Smoke Abatement Committee, 1923, Women's City Club of Cleveland Manuscript, Western Reserve Historical Society.

4. Smoke Abatement Committee, 1926, Women's City Club MSS, Western Reserve Historical Society; Regional Association of Cleveland, "Smoke Abatement Activities of the Regional Association, 1937–1941" (1941), 1–2.

5. "Dr. Moore Urges Scientific Attack on Smoke Evil," St. Louis *Post-Dispatch*, 7 March 1923; Joel Tarr and Carl Zimring, "The Struggle for Smoke Control in St. Louis: Achievement and Emulation," in *The Environmental History of St. Louis*, ed. Andrew Hurley (St. Louis: Missouri Historical Society, 1997); "A Clean Sweep for Smoke," St. Louis *Post-Dispatch*, 17 March 1923; "Smoke Injury to Park Trees," *Post-Dispatch*, 13 July 1924. For a review of smoke abatement in the early 1930s see Henry Obermeyer, *Stop That Smoke!* (New York: Harper & Brothers, 1933). Like many reformers, Obermeyer minimized the role of earlier activists, essentially dismissing the antismoke movement of the prewar decades. Obermeyer's presentation of the smoke problem was remarkably similar to that of prewar activists, with three exceptions: Obermeyer emphasized domestic smoke over industrial emissions, which may reflect both improvements in industrial abatement as well as a decline in industrial pollution due to the depression; he accepted nonvisible air pollution as a major part of the emissions problem, reflecting the growing understanding of the complexity of air pollution and the important role of invisible gases in affecting health; and he included a discussion of the role of smoke in limiting commercial air travel, which obviously was not a concern before World War I.

6. United States Bureau of Mines, "Smoke Investigation at Salt Lake City, Utah," *Bulletin* 254 (1926), 1–9.

7. H. W. Clark, "Results of Three Years of a Smoke Abatement Campaign," *American City* 31 (1924): 343–44; United States Bureau of Mines, *Bulletin* 254 (1926), 2:94–95;

Walter E. Pittman, "The Smoke Abatement Campaign in Salt Lake City, 1890–1925," *Locus* 2 (1989): 77–78.

8. John G. Clark, *Energy and the Federal Government* (Urbana: University of Illinois Press, 1987), 22; United States Bureau of the Census, *Historical Statistics of the United States* (Washington, 1960), 356, 359.

9. Chester G. Gilbert and Joseph E. Pogue, "Coal: The Resource and Its Full Utilization," *United States National Museum Bulletin* 102, pt. 4 (1917): 8.

10. Chicago *Tribune*, 31 May, 6 June, 11 July 1926, as found in Pennsylvania Railroad MSS, box 437, folder 19, Hagley Library.

11. Kurt C. Schlichting, "Grand Central Terminal and the City Beautiful in New York," *Journal of Urban History* 22 (March 1996): 332–49; Carl Condit, *The Port of New York: A History of the Rail and Terminal System from the Grand Central Electrification to the Present* (Chicago: University of Chicago Press, 1981), 190–98. See William D. Middleton, *Grand Central . . . the World's Greatest Railway Terminal* (San Marino: Golden West Books, 1977), for photographs of air rights development.

12. Chicago *Tribune*, 6 June 1926, as found in PA RR MSS, box 437, folder 19, Hagley Library.

13. Chicago *Tribune*, 6 October 1926, PA RR MSS, box 437, folder 20, Hagley Library.

14. "Proceedings of the Committee on Railway Terminals and Committee on Judiciary of the City Council of Chicago, In Re Electrification of Railway Terminals," 19 July 1926, PA RR MSS, box 438, folder 1, Hagley Library; Chicago *Evening American*, 7 August 1926; Chicago *Daily News*, 7 August 1926; Chicago *Tribune*, 12 August 1926.

15. "Statement of T. B. Hamilton, Regional Vice President," PA RR MSS, box 437, folder 15, Hagley Library; Chicago *Tribune*, 11 August 1926. George Harwood, vice president at the New York Central, suggested that updating the 1915 report would take a year. George Gibbs reported to Pennsylvania Railroad President Atterbury that he and Hamilton supported the updating project only "if the situation develops a serious turn" (Gibbs to Atterbury, 16 October 1926, PA RR MSS, box 437, folder 15, Hagley Library).

16. Sam H. Schurr and Bruce C. Netschert, *Energy in the American Economy, 1850–1975* (Baltimore: Johns Hopkins University Press, 1960), 508–9.

17. Smoke Prevention Association, "Proceedings of the 28th Annual Convention, Buffalo, New York, 1934," 30–33. At this convention Chambers read a paper written by Monnett which summarized the C.W.A. project, no doubt inspiring other cities to apply for federal funding for their own studies.

18. Frank A. Chambers to General Managers Association, 26 January 1934, PA RR MSS, box 437, folder 17, Hagley Library. For a discussion of the growing importance of radio in the lives of Chicagoans see Lizabeth Cohen, *Making a New Deal: Industrial Workers in Chicago, 1919–1939* (New York: Cambridge University Press, 1990), 129–43.

19. General Superintendent of Motive Power to F. W. Hankins, 2 February 1934,

PA RR MSS, box 437, folder 17, Hagley Library; "500 WPA Workers Analyze Air Pollution from Smoke," *Smoke* 4 (April 1939): 1; Chambers to General Managers Association, 25 January 1939, PA RR MSS, box 437, folder 13, Hagley Library.

20. Sol Pincus and Arthur Stern, "A Study of Air Pollution in New York City," *American Journal of Public Health* 27 (April 1937): 321–33; *Heating, Piping, and Air Conditioning* 17 (1945): 447–54, 495–501, 557–58.

21. Works Progress Administration, "Air Pollution, City of Pittsburgh, PA" (1940?), Smoke Investigation Activities Archive, series 1, folder 32, University of Pittsburgh.

22. City of Cleveland Department of Public Safety, "W.P.A. District No. 4, Smoke Abatement Project" (1939?); "Summary of Smoke Abatement Progress, Cleveland, Ohio: Smoke Abatement Survey, Heating, Power Plant & Incinerator Survey," (1939?).

23. "Summary of Smoke Abatement Progress, Cleveland, Ohio," 5, 22–25; Department of Municipal Research and Service, "Smoke Abatement Study of Industrial Plants and Railroads in the City of Louisville, Jefferson County, Kentucky" (1939?). The Louisville report concluded that domestic sources created more than 50 percent of the city's smoke in the heating season. See Civic Club of Allegheny County, "Smoke Memo—Meeting of Tuesday, January 28, 1941," Civic Club Records, box 11, folder 182, University of Pittsburgh.

24. William Christy, "History of the Air Pollution Control Association," *Journal of the Air Pollution Control Association* 10 (April 1960): 128–31.

25. Smoke Prevention Association, "Program 33rd Annual Convention, Milwaukee, Wisconsin, June 13–16, 1939," PA RR MSS, box 437, folder 17, Hagley Library; "Smoke, Official Bulletin Smoke Prevention Association, Inc." (May 1940); David F. Noble, *America by Design: Science, Technology, and the Rise of Corporate Capitalism* (New York: Knopf, 1977), 49. On engineers' professional status see Edwin T. Layton, *The Revolt of the Engineers: Social Responsibility and the American Engineering Profession* (Cleveland: Case Western University Press, 1971).

26. The St. Louis abatement story has received considerable attention. I have relied heavily on Joel Tarr and Carl Zimring, "The Struggle for Smoke Control in St. Louis: Achievement and Emulation," in *Common Fields*, ed. Andrew Hurley. For a more detailed accounting see Oscar Hugh Allison, "Raymond R. Tucker: The Smoke Elimination Years, 1934–1950" (Ph.D. diss., St. Louis University, 1978). Tucker himself also wrote extensively about his efforts in St. Louis: see Tucker, "A Smoke Elimination Program That Works," *Heating, Piping, and Air Conditioning* 17 (1945): 463–69, 519–23 605–10.

27. The 94 percent figure comes from the St. Louis Coal Exchange. St. Louis *Post-Dispatch*, 9 May 1939; Tarr and Zimring, "The Struggle for Smoke Control in St. Louis," 207–10. For remarks about the necessity of smokeless fuels prior to Tucker's efforts see Victor J. Azbe, "Rationalizing Smoke Abatement," in *Proceedings of the Third International Conference on Bituminous Coal*, 16–21 November 1931(Pittsburgh: Carnegie Institute of Technology, 1931), 2:593–638, esp. 637.

28. Allison, "Raymond R. Tucker," 11–15; St. Louis *Post-Dispatch*, 9 December 1936.

29. St. Louis *Post-Dispatch*, 21 January and 5, 6 February 1937.

30. St. Louis *Post-Dispatch*, 18, 22 June, 12 August, 24 November 1937; 1 December 1938; 27 March, 2 April, 23 November 1939; Tucker, "A Smoke Elimination Program That Works," 466; Allison, "Raymond R. Tucker," 20–21.

31. St. Louis *Post-Dispatch*, 26–30 November and 1–13 December 1939; *Life*, 15 January 1940, 9–11; "St. Louis Blacks-Out," *Business Week*, 13 December 1939; Tarr and Zimring, "The Struggle for Smoke Control in St. Louis," 212–14.

32. St. Louis *Post-Dispatch*, 2, 3, 12 December 1939.

33. St. Louis *Post-Dispatch*, 27 May 1939 and 25 February, 8 April 1940.

34. St. Louis *Post-Dispatch*, 16 April 1940.

35. St. Louis *Post-Dispatch*, 18 April, 26 November, 1 December 1940; Tarr and Zimring, "The Struggle for Smoke Control in St. Louis," 216–18. For praise of the ordinance see J. H. Carter, "Does Smoke Abatement Pay?" *Heating, Piping, and Air Conditioning* 18 (April 1946): 80–84.

36. Joel Tarr argues that the success that followed the 1940 efforts came from two sources: the active involvement of three key individuals, city councilman Abraham Wolk, Pittsburgh *Press* editor Edward Leech, and the Department of Health's Dr. I. Hope Alexander; and the example of St. Louis which provided a model and inspiration. See Joel Tarr and Bill Lamperes, "Changing Fuel-Use Behavior and Energy Transitions: The Pittsburgh Smoke Control Movement, 1940–1950," *Journal of Social History* 14 (Summer 1981): 561–88. I have relied heavily on Tarr's interpretation of the Pittsburgh movement.

37. Pittsburgh *Press*, 19 February 1941; Tarr and Lamperes, "Changing Fuel-Use Behavior and Energy Transitions: The Pittsburgh Smoke Control Movement," 565.

38. Civic Club of Allegheny County "Smoke Meeting," 20 February 1941 and "Report of the Mayor's Commission for the Elimination of Smoke" (1941), 1–19, Civic Club Records, box 11, folder 182, University of Pittsburgh.

39. "The Western Pennsylvania Coal Operators Association Reports on a Plan to Reduce Air Pollution in Greater Pittsburgh" (1941), 1–21, Civic Club MSS, box 11, folder 182, University of Pittsburgh.

40. Pittsburgh *Press*, 23, 24, 25 March 1941; Pittsburgh *Sun-Telegraph*, 25 March 1941, as found and labeled in Civic Club MSS, box 11, folder 189, University of Pittsburgh; "Report of the Mayor's Commission," 8–11.

41. Tarr and Lamperes, "Changing Fuel-Use Behavior and Energy Transitions," 565–71; "Report of the Mayor's Commission," 19.

42. Tarr and Lamperes, "Changing Fuel-Use Behavior and Energy Transitions," 572–75; "Minutes Special Meeting of the United Smoke Council of the Allegheny Conference," 23 February 1946, Civic Club MSS, box 14, folder 220, University of Pittsburgh.

43. Tarr and Lamperes, "Changing Fuel-Use Behavior and Energy Transitions," 574–75; United Smoke Council, "That New Look in Pittsburgh" (1948?), Civic Club MSS, box 14, folder 221, University of Pittsburgh.

44. Tarr and Lamperes, "Changing Fuel-Use Behavior and Energy Transitions," 576; Joel Tarr, "Railroad Smoke Control: The Regulation of a Mobile Pollution Source," in *The Search for the Ultimate Sink* (Akron: University of Akron Press, 1996), 277–78.

45. Smoke Abatement League, "1928, Annual Report of the Superintendent," 2; "1933, Annual Report of the Superintendent," 1–4; Smoke Abatement League, "Smoke and Soot Pollution Analysis, 1934–1935" (1935), 3.

46. Smoke Abatement League, "1937, Annual Report," 3.

47. Smoke Abatement League, "1941, Annual Report," 1.

48. Taft, son of William Howard Taft, was the nephew of the Charles P. Taft who had published the *Times-Star* and had been so active in the pre–World War I antismoke movement.

49. Gore to Leggett and Company, 4 April 1940; Border to Taft, 5 April 1940; Taft to Kreger and Leggett, 8 April 1940, all in Charles P. Taft MSS, box 32, folder 10, Cincinnati Historical Society.

50. Castellini to Taft, 2 April 1940; Harold M. Buzek to Taft, 2 April 1940; Taft to Buzek, 3 April 1940, all in Taft MSS, box 32, folder 10, Cincinnati Historical Society.

51. For brief histories of the Coal Producers Committee for Smoke Abatement see H. B. Lammers, "Abatement of Smoke Can Make Markets for Virginia Coal" (1949), and "Report of Annual Meeting of Coal Producers Committee for Smoke Abatement" (Cincinnati, 1951), 4–9, both found in Westmoreland Coal Archive, box 416, folder 7, Hagley Library.

52. Metropolitan Smoke Control Committee, *Testimony Presented to Cincinnati City Council Law Committee* (Metropolitan Smoke Control Committee, 1942), 2–67. The committee was chaired by Julian E. Tobey of Appalachian Coals, Inc. Tobey had been interested in the smoke issue for several years. In 1938 he spoke before the Smoke Prevention Association on "The Practical Significance of Smoke Abatement." He argued that Appalachian Coals was working hard to abate smoke, noting his company's organization of a smoke abatement conference in Cincinnati in 1936. Tobey most likely hoped his presence might offset that of Tucker, who also spoke before the association. See Smoke Prevention Association, *Manual of Ordinances and Requirements in the Interest of Air Pollution Smoke Elimination and Fuel Combustion* (1938).

53. Metropolitan Smoke Control Committee, 2–3.

54. Cincinnati *Times-Star*, 6 January 1942; Clarence A. Mills, *Air Pollution and Community Health* (Boston: Christopher Publishing House, 1954), 90–106.

55. Smoke Abatement League, "1942, Annual Report," 1–5.

56. Cincinnati had a three-party system, with the Charter Party joining the national parties. Charterite Councilman Albert Cash, who later served as mayor from 1948 to 1951, had introduced the St. Louis-type bill in 1941.

57. Charles P. Taft, "Address before the meeting of the City Charter Committee, December 4, 1946," Taft MSS, box 4, folder 2, Cincinnati Historical Society; *City Bulletin*, 11 March 1947, 6–10; Air Pollution Control League, "Twenty-Sixth Annual Soot & Dustfall Report," 10 October 1956.

58. Arthur C. Stern, "General Atmospheric Pollution," *American Journal of Public Health* 38 (July 1948): 966–69.

59. New York *Times*, 3, 8, 17 January 1947.

60. New York *Times*, 7 March 1947 and 5 May, 8 June 1948.

61. New York *Times*, 3, 15 June, 21 September, 28 November 1948; Winfield Scott Downs, ed., *Who's Who in New York (City and State), 1947* (New York: Lewis Historical Publishing Company, 1947). In his 1997 doctoral dissertation, Scott Hamilton Dewey offers a fuller description of the postwar movement, based largely on New York *Times* articles. Unfortunately, Dewey inexplicably splits the story into two chapters, one concerning bureaucratic changes and the other public pressure. This division makes for an unnecessarily awkward telling, particularly since the bureaucratic story precedes his discussion of public activism. See Scott Hamilton Dewey, "'Don't Breathe the Air': Air Pollution and the Evolution of Environmental Policy and Politics in the United States, 1945–1970" (Ph.D. diss., Rice University, 1997).

62. New York *Times*, 22, 31 October and 1, 2, 20 November 1948; Lynne Page Snyder, "'The Death-Dealing Smog over Donora, Pennsylvania': Industrial Air Pollution, Public Health Policy, and the Politics of Expertise, 1948–1949," *Environmental History Review* 18 (Spring 1994): 117–39.

63. New York *Times*, 2, 3 February, 1 March, 7 July, 16 August, 28 October 1949. Dewey reaches a different conclusion from his reading of the *Times* articles, claiming that Donora "stimulated action by the administration in New York City." His conclusion seems to be based largely on the fact that the antismoke bill passed four months after Donora, which given the frequency of new smoke laws (three in five years) seems thin evidence. See Dewey," 'Don't Breathe the Air,'" 322–24. Christy had been active in the Smoke Prevention Association since 1931, representing Hudson County, New Jersey, where he headed the nation's only county-based smoke department. Christy wrote a brief history for the organization in 1960: "History of the Air Pollution Control Association," *Journal of the Air Pollution Control Association* 10 (1960): 126–37.

64. New York *Times*, 13 August 1949 and 27 June, 16 July, 4, 9, 14 August 1950.

65. New York *Times*, 19 August, 20 September, 1 October, 15 December 1950 and 3 January 1951.

66. New York *Times*, 11, 17 January, 1 February 1951. As the *Times* later made public, Robinson received financial support from anthracite concerns as she started the committee. She later returned the money (2 August 1951).

67. New York *Times*, 2 February, 9, 20 May 1951; Committee for Smoke Control, "What We Breathe" (New York, 1951).

68. New York *Times*, 6 December 1951 and 24 June, 25 September, 16 November 1952.

69. Coal Producers Committee for Smoke Abatement, "A Survey of Heating and Power Plants, City of Cleveland, with Recommendations for the Elimination of Smoke" (Cincinnati, 1946); Herbert G. Dyktor, "General Atmospheric Pollution,"

American Journal of Public Health 38 (July 1948): 957–59; Snyder, "'The Death-Dealing Smog over Donora, Pennsylvania,'" 126.

70. Christopher C. Sellers, *Hazards of the Job: From Industrial Disease to Environmental Health Science* (Chapel Hill: University of North Carolina Press, 1997), 208–24, 236. Kehoe quotation as found in Sellers, 224.

71. City of Cleveland Division of Air Pollution Control Bureau of Smoke Abatement, "Annual Report for 1947," 1–3; Tarr and Lamperes, "Changing Fuel-Use Behavior and Energy Transitions," 576–81.

conclusion The Struggle for Civilized Air

1. Francis B. Crocker, "Coalless Cities," *Cassier's Magazine* 9 (1896): 231–38; "How Chicago's Smoke Problem Is Gradually Solving Itself," Chicago *Record-Herald*, 21 November 1908; Bureau of Mines, *Information Circular 7016* (1938), 9. Martha Bruere also called electricity "A Cure for Smoke-Sick Cities," *Collier's* 174 (1924): 28. Electricity, she thought, would make cities cleaner and less congested.

2. Petroleum-derived diesel fuel was slow to replace coal on locomotives, with mass conversions to diesel-electric technology awaiting the 1940s. Thomas G. Marx, "Technological Change and the Theory of the Firm: The American Locomotive Industry, 1920–1955," *Business History Review* 50 (1976): 7–9; Joseph A. Pratt, "The Ascent of Oil: The Transition from Coal to Oil in Early Twentieth-Century America," in *Energy Transitions: Long-Term Perspectives*, ed. Lewis J. Perelman (Boulder: Westview Press, 1981), 11–22. When New York and other eastern cities struggled through the coal shortages brought on by the 1902 anthracite strike, oil seemed a potential alternative to coal. With discoveries of oil in Texas and California in 1900 and 1901, greater supplies of higher quality petroleum began to reach urban markets, and with the higher coal prices during the strike, petroleum suddenly presented a viable alternative. "Oil May Supplant Coal," announced the New York *Tribune*; "Use Oil in Place of Coal," read the New York *Times*. But the coal shortage brought no great shift to fuel oil, particularly when cheap coal prices returned after the strike (New York *Tribune*, 6, 12 June 1902; New York *Times*, 17 June 1902).

3. John H. Herbert, *Clean, Cheap Heat* (New York: Praeger, 1992), 35. The nation's most popular natural gas trade magazine lobbied for gas interests to become involved in the national antismoke campaign. One editorial asserted, "[T]he smoke nuisance is a splendid argument to use in selling gas. . . . What we need is a serious appreciation of this opportunity and then a campaign, more persistent than demonstrative, but in dead earnest" ("Smoke Nuisance," *Progressive Age* 26 [1908]: 543–44). See also Mark H. Rose, *Cities of Light and Heat: Domesticating Gas and Electricity in Urban America* (University Park: Pennsylvania State University Press, 1995). Rose's work focuses on Kansas City and Denver, two cities with greater access to the natural gas fields of the southern plains.

4. For a discussion of the benefits of electrification for Chicago businesses see

Harold Platt, *The Electric City* (Chicago: University of Chicago Press, 1991), 208–20. Some electric power companies faced public scorn for producing smoke at their generating stations. In New York, where Consolidated Edison operated large coal-fired plants in Manhattan, the power company became a central target of antismoke agitation and the city's control effort in the early 1950s (New York *Times*, 22, 23 June; 4, 9 August; 19, 24 September 1950).

5. "Smokeless Cities," New York *Tribune*, 26 December 1901. Although electricity had much to recommend it as a superior alternative to coal power, electric power companies did emphasize the relative healthfulness and cleanliness of electricity as compared with coal. "Electric service is the most healthful and economical of modern conveniences," read one pamphlet published by the Milwaukee Electric Railway and Light Company. "Neither in lighting, cooking, heating, nor through any of its mechanical devices does it destroy any healthful element of the air we breath." The pamphlet stated more directly, "It does not scatter dust, with its disease germs, through the atmosphere nor fill the house with odors or greasy vapors" ("The Electrical House That Jack Built" [Milwaukee, 1916]).

6. Harold Platt, *The Electrical City: Energy and the Growth of the Chicago Area, 1880–1930* (Chicago: University of Chicago Press, 1991), 208–20; Sam H. Schurr and Bruce C. Netschert, *Energy in the American Economy, 1850–1975* (Baltimore: Johns Hopkins Press, 1960), 81–83.

7. Joel Tarr and Bill Lamperes, "Changing Fuel-Use Behavior and Energy Transitions: The Pittsburgh Smoke Control Movement, 1940–1950," *Journal of Social History* 14 (Summer 1981): 561–88; Schurr and Netschert, *Energy in the American Economy*, 83; Joel Tarr and Kenneth Koons, "Railroad Smoke Control: A Case Study in the Regulation of a Mobile Pollution Source," in *Energy and Transport: Historical Perspectives on Policy Issues*, ed. George H. Daniels and Mark H. Rose (Beverly Hills: Sage Publications, 1982), 71–92; Thomas G. Marx, "Technological Change and the Theory of the Firm: The American Locomotive Industry, 1920–1955," *Business History Review* 50 (Spring 1976): 1–24. For a discussion of coal industry competition see James P. Johnson, *The Politics of Soft Coal* (Urbana: University of Illinois Press, 1979), 95–134.

8. H. B. Lammers, "Abatement of Smoke Can Make Markets for Virginia Coal," delivered before the Virginia Coal Operators Association annual meeting, June 1949, p. 2, in Westmoreland Coal MSS, box 416, folder 7, Hagley Library; Samuel Hays, *Beauty, Health, and Permanence: Environmental Politics in the United States, 1955–1985* (New York: Cambridge University Press, 1987).

9. Henry Obermeyer, *Stop That Smoke!* (New York: Harper & Brothers, 1933), 233; Lammers, "Abatement of Smoke Can Make Markets for Virginia Coal," 4.

10. Railroad consumption of coal fell from 128 million tons in 1945 to 3 million tons in 1960. Retail sales, directed mostly at domestic users, fell from 119 million tons to 30 million in the same years. By 1970, electric utilities accounted for over 60 percent of the nation's coal consumption (Senate Committee on Government Affairs, *The Coal Industry: Problems and Prospects. A Background Study*, 95th Cong., 2nd sess., 1978, 48–49).

11. William G. Christy, "History of the Air Pollution Control Association," *Journal of the Air Pollution Control Association* 10 (1960): 134–35; Coal Producers Committee for Smoke Abatement, "Review," May 1954, Westmoreland MSS, box 416, folder 7, Hagley Library.

12. Clive Howard, "Smoke: The Silent Murderer," *Woman's Home Companion* 76 (February 1949): 332–33+; Clarence A. Mills, *Air Pollution and Community Health* (Boston: Christopher Publishing House, 1954), 81. Many Americans also read about similar episodes outside the United States, including the story of a hydrogen sulfide leak at a natural gas refinery at Poza Rica, Mexico, which killed twenty-two people (Louis C. McCabe, "Atmospheric Pollution," *Industrial and Engineering Chemistry* 43 [February 1951]: 79A).

13. Los Angeles *Times*, 19 January 1947. For the story of Los Angeles's early struggles against smog see: Marvin Brienes, "The Fight against Smog in Los Angeles, 1943–1957" (Ph.D. diss., University of California–Davis, 1975); James E. Krier and Edmund Ursin, *Pollution and Policy: A Case Essay on California and Federal Experience with Motor Vehicle Air Pollution, 1940–1975* (Berkeley: University of California Press, 1977); Scott Hamilton Dewey, "'Don't Breathe the Air': Air Pollution and the Evolution of Environmental Policy and Politics in the United States, 1945–1970" (Ph.D. diss., Rice University, 1997), 128–315.

14. Lose Angeles *Times*, 19 January 1947; Oscar Hugh Allison, "Raymond Tucker: The Smoke Elimination Years, 1934–1950" (Ph.D. diss., St. Louis University, 1978), 173–79.

15. Ronald Schiller, "The Los Angeles Smog," *National Municipal Review* 44 (1955): 558–64. For a brief discussion of the role of World War II in changing the nature of L.A.'s economy see Frank M. Stead, "Study and Control of Industrial Atmospheric Pollution Nuisances," *American Journal of Public Health* 35 (1945): 491–98.

16. Robert Dale Grinder, "The Battle for Clean Air," in *Pollution and Reform in American Cities, 1870–1930*, ed. Martin Melosi (Austin: University of Texas Press, 1980), 101. Grinder draws the same conclusion in his dissertation, "The Anti-Smoke Crusades" (Ph.D. diss., University of Missouri, 1973). Grinder is hardly alone in underestimating the effect of progressive-era antismoke activists. Political scientist Charles O. Jones, in his well-known work on air pollution, dismisses prewar antismoke efforts, even declaring the very thorough Mellon investigation "hardly sufficient to serve as a basis for comprehensive legislation." Jones also inexplicably asserts that the smoke issue did not reach city agendas because of weak problem identification, ignoring the fact that the problem was well defined, that antismoke groups organized around the nation, and that activists influenced public policy and indeed changed the shape of municipal government. See Charles O. Jones, *Clean Air* (Pittsburgh: University of Pittsburgh Press, 1975), 22–23. Another scholar, Matthew Crenson, based an entire study on the premise that cities did not act on air pollution until after World War II, and in some cases not until long after the war. See Crenson, *The Un-Politics of Air Pollution* (Baltimore: Johns Hopkins University Press, 1971). Scott Dewey ("'Don't

Breathe the Air'") not only concludes that the progressive-era movements failed, based largely on his reading of Grinder, but also that the antismoke efforts of the 1940s and 1950s largely failed as well. Unfortunately, Dewey does not distinguish smoke from "air pollution."

17. Unfortunately, no reliable scientific data concerning air quality exists for these decades. See estimates of improved air quality in Cincinnati and New York in chapter 3, and in Philadelphia, Pittsburgh, and Rochester in chapter 4. For estimates of smoke in Pittsburgh dating back to the 1840s see Cliff Davidson, "Air Pollution in Pittsburgh: A Historical Perspective," *Journal of the Air Pollution Control Association* 29 (1979): 1035–41.

18. This total reflects the addition of Brooklyn's 566,000 residents, who were not yet New York City residents, but who would be in 1920.

19. Bureau of the Census, *Fourteenth Census of the United States* (1920), v. 1, 80–81, 50; Bureau of the Census, *Seventeenth Census of the United States* (1950), v. 1, 5. In 1950 the census began to use a new rubric for calculating urban populations which allowed the inclusion of more distant suburban populations. The census also included the figures as they would have been determined under the old rubric. I have used those numbers.

20. Schurr and Netschert, *Energy in the American Economy,* 508.

21. "Progress and Possibilities in the Abatement of Smoke," *The American City* 43 (1930): 125.

Bibliographical Essay

PRIMARY SOURCES

Manuscript Collections

The Mellon Institute Smoke Investigation Activities manuscript contains the most valuable collection of documents relating to the progressive-era antismoke movements. Held at the University of Pittsburgh's Archives of Industrial Society, this collection includes rich documentation of the Mellon Institute's original smoke investigation and subsequent research in the 1920s and 1930s. It also contains many smoke-related documents from around the nation and Britain, which Mellon fellows collected as they conducted their research. Allegheny County's Civic Club Records also reside in the University of Pittsburgh collection. This manuscript contains valuable information concerning that city's 1940s antismoke movement, including documents from the United Smoke Council.

I also found to be extremely helpful the Pennsylvania Railroad manuscript at the Hagley Library, Wilmington, Delaware. Well organized and indexed, these records include considerable information concerning the electrification debates in Chicago, especially for the dates 1907–17 and 1926–28, as well as antismoke agitation in several other cities, including Washington, Philadelphia, and Cincinnati. The manuscript also contains smoke-related documents of a more technical nature, particularly from the railroad's research department at Altoona. Also at the Hagley, the Westmoreland Coal Company manuscript contains several smoke-related files which hold documents related to the Coal Producers Committee for Smoke Abatement.

I conducted much of my early research in Cincinnati, where the Cincinnati Historical Society holds a wealth of information from the Smoke Abatement League of Hamilton County. The Charles P. Taft Papers also contain files of interest concerning that city's efforts in the 1940s. Also in Cincinnati, I was fortunate to receive access to the Woman's Club records, which are held in private by the club. These documents provided great insight into the philosophy that guided women in their crusade.

Cleveland's Western Reserve Historical Society holds the records of the Women's City Club of Cleveland, which contain detailed documentation of that organization's efforts in the 1920s.

Newspapers

As this study was largely concerned with the movement to abate smoke, published material was particularly important. The coverage in newspapers and magazines gave important clues as to the development of the problem and served as a barometer of the issue's salience. Several libraries contain valuable collections of newspaper clippings. The Cincinnati Historical Society holds the Smoke Abatement League's scrapbook for the years 1906–1919. On microfilm and organized chronologically, this scrapbook contains hundreds of clippings from the *Cincinnati Enquirer, Post, Times-Star, Citizen's Bulletin, Commercial Tribune, Western Architect and Builder, Western Christian Advocate,* and smaller numbers of clippings from several out-of-town papers. The Cleveland Public Library also contains a scrapbook on the subject, with clippings dating from 1900 through 1909. Those articles come from the *Cleveland News, Plain Dealer,* and *Press,* as well as many out-of-town papers, including the *Indianapolis News* and the *Milwaukee Sentinel.* The Milwaukee Municipal Reference Library contains a microfilmed "Smoke Control Clippings" collection, with articles from the *Milwaukee Evening Wisconsin, Free Press, Journal, Leader, and Sentinel* for the 1910s and 1920s. Finally, the Pittsburgh Carnegie Library holds a "Smoke Control Collection" in its Pennsylvania Room, with clippings from the *Pittsburgh Gazette, Post, Civic News,* and others. These clippings date from the 1910s through the 1940s.

In addition, indexes make several other papers valuable, including the *Chicago Record-Herald,* 1904–1912, *Cincinnati Enquirer,* 1930–1955, *Cincinnati Post,* 1930–1955, *Cleveland Leader,* 1869–1876, *Milwaukee Sentinel,* 1879–1890, *New York Tribune,* 1892–1902, and *New York Times,* 1892–1960. Given the important role of the press in the antismoke movement, I also searched through several unindexed papers, including the *Birmingham Age-Herald,* 1913, *Birmingham Labor-Advocate,* 1913, *Chicago Tribune,* 1913–15, and *Los Angeles Times,* 1946.

Periodicals

Surprisingly, perhaps the most valuable periodical source for progressive-era smoke abatement information is the little-known *Industrial World,* 1907–14, and its successor *Steel and Iron,* 1914–15. These Pittsburgh-based steel industry magazines contain a wealth of information concerning the International Association for the Prevention of Smoke, Bureau of Mines research, and the Mellon investigation. The engineering press provides abundant information concerning the technical aspects of smoke abatement and unexpectedly thorough coverage of politics. Most valuable are *Power, Iron Age, Engineering News, Engineering Record, Engineering Magazine, Journal of the Franklin Institute, Scientific American, Cassier's,* and *Heating, Piping, and Air Conditioning.* Also valuable are the published records of several engineering societies, including the *Transactions of the American Society of Civil Engineers,* the *Transactions of the American Society of Mechanical Engineers,* and the *Engineers Society of Western Pennsylvania Proceedings.* The railroad press also contains thorough coverage of the smoke and electrification issues. Particularly helpful are *Railroad (Railway) Age* and *Electric Railway Journal*. Several engineering schools also published bulletins, some of which provided important detail on academic abatement research. Most important among these bulletins is L. P. Breckenridge, "How to Burn Illinois Coal without Smoke," *University of Illinois Engineering Experiment Station Bulletin Number 15* (August 1907). See also Ernest L. Ohle and Leroy McMaster, "Soot-Fall Studies in Saint Louis," *Washington University Studies* 5, pt. 1 (1917): 3–8.

Medical journals also contain articles concerning smoke. Although much later and much less than Britain's *Lancet*, the *Journal of the American Medical Association* did discuss the health effects of smoke, as did *Medical News* and *Medical Record*. Chicago's *Sanitary News*, with a short run in the late 1800s, published dozens of articles on Chicago's early efforts to control smoke and provided insight into the reasoning behind early abatement efforts. In the 1930s and 1940s, the *American Journal of Public Health* ran more than a dozen important articles concerning air pollution and health.

Finally, the reform-minded magazines *American City* and *National Municipal Review* followed the issue closely. Several influential articles appeared in the popular press, including James Parton's essay on Pittsburgh in the *Atlantic Monthly* (21 [1868]: 17–28). For the most part, however, other than newspapers the popular press offered only fleeting coverage of the smoke problem, and was valuable only in determining the relative importance of smoke abatement among the numerous issues of the day.

Government Documents

The United States Bureau of Mines and the Geological Survey published a number of important smoke-related documents after 1900. These bulletins, circulars, and professional papers provide both a full accounting of the federal involvement in smoke research, and several offer fine descriptions of the state of antismoke engineering. From the Geological Survey see Edward Parker, Joseph Holmes, and Marius Campbell, "Report on the Operations of the Coal-Testing Plant of the United States Geological Survey at the Louisiana Purchase Exposition, St. Louis, MO., 1904," U.S.G.S. Professional Paper 48 (1906); D. T. Randall, "The Burning of Coal without Smoke in Boiler Plants: A Preliminary Report," U.S.G.S. Bulletin 334 (1908); and Randall and H. W. Weeks, "The Smokeless Combustion of Coal in Boiler Plants," U.S.G.S. Bulletin 373 (1909). From the Bureau of Mines see Samuel B. Flagg, "Smoke Abatement and City Smoke Ordinances," Bureau of Mines Bulletin 49 (1912); Osborn Monnett, "Smoke Abatement," Bureau of Mines Technical Paper 273 (1923); Monnett, G. J. Perrott, and H. W. Clark, "Smoke Abatement Investigation at Salt Lake City, Utah," Bureau of Mines Bulletin 254 (1926).

I also found several other government documents of value. Harry Garfield's *Final Report of the United States Fuel Administrator, 1917–1919* (1921) contains a fine description of the government's fuel policies during the war, including conservation efforts. Chester Gilbert published an extensive report on coal through the United States National Museum, Bulletin 102 (1917), which includes some wonderful statistics and ruminations on the fuel. The Smithsonian Institution also published some important papers concerning smoke, particularly J. B. Cohen's "The Air of Towns," *Annual Report of the Board of Regents of the Smithsonian Institution* (1913), 653–85.

Local governments produced a wealth of documentation on the smoke abatement effort. Annual reports issued by smoke departments exist for most large cities, generally dating from the early 1910s. Although other libraries do hold collections of such reports, municipal reference libraries contain the most complete runs. Most important for this study were the following: the Cincinnati smoke inspector's reports, which were published in the city's *Annual Reports of the Departments;* Milwaukee's "Annual Report of the Smoke Inspector"; and Philadelphia's Smoke Department Annual Reports, also published with reports from the other city departments. Health department annual reports were also valuable sources of information and provided an indication of the level of concern for smoke as a health issue. I used the reports from Cincinnati and New York most thoroughly. Many cities also published W.P.A. studies during the latter half of the 1930s, including Pittsburgh, Cleveland, and Louisville. These reports consist mainly of minute detail, but also contain thorough descriptions of the W.P.A. projects' staffs, goals, and tactics.

Other Primary Sources

Two contemporary smoke studies provide great detail on the scope and nature of the problem. First, and most important, the Mellon Smoke Investigation, published in nine bulletins over two years, contains rich material on every aspect of the issue, including health, economics, engineering, weather, and vegetation. The study also issued a lengthy and helpful bibliography as its second bulletin. Published by the University of Pittsburgh in 1913 and 1914, these bulletins are usually bound together as *Mellon Institute of Industrial Research Smoke Investigation.* Second, the Chicago Association of Commerce's massive report on electrification provides fine detail concerning Chicago's problem and insight into the corporate response to antismoke agitation. See Chicago Association of Commerce, *Smoke Abatement and Electrification of Railway Terminals in Chicago* (Chicago, 1915).

For the legal aspect of smoke abatement see Horace Gay Wood, *A Practical Treatise on the Law of Nuisances in Their Various Forms* (Albany: John D. Parsons, Jr., Publisher, 1875). For citations to and descriptions of more recent cases see Jan G. Laitos, "Continuities from the Past Affecting Resource Use and Conservation Patterns," *Oklahoma Law Review* 28 (Winter 1975): 60–96, and Harold W. Kennedy and Andrew O. Porter, "Air Pollution: Its Control and Abatement," *Vanderbilt Law Review* 8 (1955): 854–77. Of course, there is no substitution for the court reporters themselves, which I consulted at great length. Also valuable is Lucius H. Cannon, *Smoke Abatement: A Study of the Police Power as Embodied in Laws, Ordinances, and Court Decisions* (St. Louis: St.

Louis Public Library, 1924). A compendium of court cases and laws from dozens of cities, states, and countries, Cannon's work provides important detail about places that I could not visit and pointed me toward relevant court decisions.

Progressive-era public interest groups produced a great deal of information on the smoke issue. Most important for this study, Cincinnati's Smoke Abatement League published annual reports from the 1910s through the 1950s, although not all of them are available. These reports not only contain summaries of events from the year but also provide insight into the reform philosophy. The Cleveland Chamber of Commerce published a series of reports from their Committee on Smoke Prevention (1907, 1909, 1912, 1914). I also found dozens of pamphlets of interest, including: Citizens' Association of Chicago, "Report of the Smoke Committee" (May 1889); Cleveland Regional Association, "Smoke Abatement Activities of the Regional Association, 1937–1941" (1941); Robert H. Fernald, "The Smoke Nuisance," University of Pennsylvania Free Public Lecture Course (1915); St. Louis Civic League, "Report of the Smoke Abatement Committee," (November 1906); Samuel W. Skinner and William H. Bryan, "Smoke Abatement," papers read before the Optimist Club (Cincinnati, 1899); Syracuse Chamber of Commerce, "Report upon Smoke Abatement" (1907); and Elliot H. Whitlock, "Watch Our Smoke" (Cleveland Chamber of Commerce, 1926).

The Cincinnati Historical Association also holds two published addresses by Charles A. L. Reed, both of which provide wonderful insight into the reform philosophy: "An Address on the Smoke Problem," delivered before the Woman's Club of Cincinnati (April 24, 1905); "The Smoke Campaign in Cincinnati," remarks made before the National Association of Stationary Engineers (July 10, 1906).

Several contemporary monographs concerning smoke and the city are also of value, including the following: John H. Griscom, *The Uses and Abuses of Air: Showing Its Influences in Sustaining Life, and Producing Disease* (New York: J. S. Redfield, 1848); George Derby, *An Inquiry into the Influence of Anthracite Fires upon Health* (Boston: A. Williams & Co., 1868); and, Joseph W. Hays, *Combustion and Smokeless Furnaces* (New York: Hill Publishing, 1906). Henry Obermeyer's *Stop That Smoke!* (New York: Harper & Brothers, 1933) offers a fine summary of the state of the knowledge about smoke in the early 1930s, and provides insight into the major concerns for antismoke activists. For a progressive reformers' thoughts on urbanism, see Frederic C. Howe, *The City: The Hope of Democracy* (Seattle: University of Washington Press, 1967; reprint of 1905 original), and the several works by Charles Mulford Robinson, including *The Improvement of Towns and Cities* (New York: Putnam's, 1901), and *City Planning* (New York: Putnam's, 1916).

I also read contemporary fiction in search of smoke symbolism, which I found in abundance. See especially Charles Dickens, *Hard Times* (New York: Oxford University Press, 1992; orig. pub. 1854); Theodore Dreiser, *Sister Carrie* (New York: World Publishing, 1951; orig. pub. 1900); Hamlin Garland, *Rose of Dutcher's Coolly* (Chicago: Stone & Kimball, 1895); Robert Herrick, *Waste* (New York: Harcourt, Brace, 1924);

Frank Norris, *The Pit: A Story of Chicago* (New York: Sun Dial Press, 1937; orig. pub. 1903); Upton Sinclair, *The Jungle* (New York: Signet New American Library, 1960; orig. pub. 1906); and Booth Tarkington, *Growth* (New York: Doubleday, Page, 1927).

SECONDARY SOURCES

Like any environmental history, this study relies on the work of historians working in many different fields, from political history to labor history, from the history of medicine to the history of technology. I have categorized these secondary works for easier reference.

Smoke and Air Pollution

Joel Tarr's many articles on the urban environment have provided important guidance for my work. Now collected in one volume, *The Search for the Ultimate Sink* (Akron, Ohio: University of Akron Press, 1996), these articles concern several important topics, including Pittsburgh's 1940s smoke control movement, railroad smoke control, and the Pittsburgh Survey. Another important Tarr essay, written with Carl Zimring, on the St. Louis movement of the 1930s and 1940s, appears in a recent collection edited by Andrew Hurley, *Common Fields: An Environmental History of St. Louis* (St. Louis: Missouri Historical Society Press, 1997). Also valuable is his article concerning pollution in the iron industry, "Searching for a 'Sink' for an Industrial Waste: Iron-Making Fuels and the Environment," *Environmental History Review* 18 (1994): 9–34. In addition, Tarr has written on other environmental issues, including several articles on water quality, which appear in *The Search for the Ultimate Sink*. See also his work with Mark Tebeau, "Managing Danger in the Home Environment, 1900–1940," *Journal of Social History* 29 (1996): 797–816. These articles provided valuable context for my study.

Robert Dale Grinder produced the first dissertation on the smoke abatement movement, "The Anti-Smoke Crusades: Early Attempts to Reform the Urban Environment, 1883–1918" (Ph.D. diss., University of Missouri–Columbia, 1973. Grinder also published two articles on the issue: "The War against St. Louis's Smoke, 1891–1924," *Missouri Historical Review* 69 (1975): 191–205; and "From Insurgency to Efficiency: The Smoke Abatement Campaign in Pittsburgh before World War I," *The Western Pennsylvania Historical Magazine* 61 (1978): 187–202. Until now, Grinder's chapter in Martin Melosi's *Pollution and Reform in American Cities, 1870–1930* (Austin: University of Texas Press, 1980) has provided the most accessible analysis of the progressive-era movement. Grinder's work is insightful, and his bibliography is helpful.

Several other historians have written on the subject of smoke control. See Harold Platt's inexplicably titled "Invisible Gases: Smoke, Gender, and the Redefinition of Environmental Policy in Chicago, 1900–1920," *Planning Perspectives* 10 (1995): 67–97. Platt's primary concern with germ theory and his failure to place Chicago in a national context lead him to dubious conclusions, but his article guided me to an important story and the sources that tell it. Also concerning Chicago see Christine Meisner Rosen, "Businessmen against Pollution in Late Nineteenth Century Chicago," *Business History Review* 69 (Autumn 1995): 351–397. Rosen's work provides analysis of the effort to clear the air for the Columbian Exposition. Walter E. Pittman's "The Smoke Abatement Campaign in Salt Lake City, 1890–1925," *Locus* 2 (1989): 69–78, pointed me out of the Midwest and East. For more on western smoke problems see Donald MacMillan, "A History of the Struggle to Abate Air Pollution from Copper Smelters of the Far West, 1885–1933" (Ph.D. diss., University of Montana, 1973). Also of interest is John Duffy, "Smoke, Smog, and Health in Early Pittsburgh," *Western Pennsylvania Historical Magazine* 45 (1962): 93–106; and Harold C. Livesay and Glenn Porter, "William Savery and the Wonderful Parsons Smoke-Eating Machine," *Delaware History* 14 (1971): 161–76.

British smoke has also received considerable attention from historians, and much of that work is valuable for understanding the context of the American problem. Two book-length studies exist: Peter Brimblecombe, *The Big Smoke: A History of Air Pollution in London since Medieval Times* (New York: Methuen, 1987); and Eric Ashby and Mary Anderson, *The Politics of Clean Air* (New York: Oxford University Press, 1981). Brimblecombe also produced a valuable article, "Attitudes and Responses towards Air Pollution in Medieval England," *Journal of the Air Pollution Control Association* 26 (October 1976): 941–45. I also found Peter Thorshiem's work, "Air Pollution and Anxiety in Late Nineteenth-Century London" (M.A. thesis, University of Wisconsin–Madison, 1994) to be valuable, as were our many conversations concerning smoke on both sides of the Atlantic.

Too little historical analysis of the later air pollution control efforts exists. Much of what we have is the work of political scientists interested in the governmental response, or failed response, to a scientifically and politically complex issue. While completely ignoring early antismoke efforts, Matthew Crenson's *The Un-Politics of Air Pollution: A Study of Non-Decisionmaking in the Cities* (Baltimore: Johns Hopkins University Press, 1971) is of some value. More valuable is Charles O. Jones, *Clean Air: The Policies and Politics of Pollution Control* (Pittsburgh: University of Pittsburgh Press, 1975). On Los Angeles see Marvin Brienes, "The Fight against Smog in Los Angeles, 1943–1957" (Ph.D. diss., University of California–Davis, 1975); and James E. Krier and Edmund Ursin, *Pollution and Policy: A Case Essay on California and Federal Experience with Motor Vehicle Air Pollution, 1940–1975* (Berkeley: University of California Press, 1977). For more recent work see Scott Hamilton Dewey, "'Don't Breathe the Air: Air Pollution and the Evolution of Environmental Policy and Politics in the United States, 1945–1970" (Ph.D. diss., Rice University, 1997). Dewey tells the stories of Los Angeles (to which he adds little) and New York, but he also includes interesting chapters on rural air pollution in central Florida. Although her work has a narrow focus, Lynne Page Snyder's "'The Death-Dealing Smog over Donora, Pennsylvania': Industrial Air Pollution, Public Health Policy, and the Politics of Expertise, 1948–1949," *Environ-*

mental History Review 18 (Spring 1994): 117–39, provides an excellent analysis of an important event.

Health and Medicine

Several historical studies of public health and medicine provided context for this study. Perhaps most important is John Duffy, *A History of Public Health in New York City, 1866–1966* (New York: Russell Sage Foundation, 1974). See also Judith Leavitt, *The Healthiest City: Milwaukee and the Politics of Health Reform* (Princeton: Princeton University Press, 1982). For a unique depiction of the changing ideas concerning infectious disease see Charles Rosenberg, *The Cholera Years: The United States in 1832, 1849, and 1866* (Chicago: University of Chicago Press, 1962). Nancy Tombs's recent article on germ theory has been very influential and has certainly affected this work. See "The Private Side of Public Health: Sanitary Science, Domestic Hygiene, and Germ Theory, 1870–1900," *Bulletin of the History of Medicine* 64 (1990): 509–39. On tuberculosis specifically, see Barbara Bates, *Bargaining for Life: A Social History of Tuberculosis, 1876–1938* (Philadelphia: University of Pennsylvania Press, 1992); and Michael E. Teller, *The Tuberculosis Movement: A Public Health Campaign in the Progressive Era* (New York: Greenwood Press, 1988).

Energy

I relied very heavily on the work of Sam H. Schurr and Bruce C. Netschert, whose *Energy in the American Economy, 1850–1975* (Baltimore: Johns Hopkins University Press, 1960) contains extensive statistical data on the several fuel transitions that occurred over the years of their study (and this study). For more analysis of the role of energy in American society see Martin Melosi, *Coping with Abundance: Energy and Environment in Industrial America* (Philadelphia: Temple University Press, 1985). Also valuable is Melosi's chapter, "Energy Transitions in the Nineteenth-Century Economy," in *Energy in Transport: Historical Perspectives on Policy Issues*, ed. George H. Daniels and Mark H. Rose (Beverly Hills: Sage Publications, 1982). See also Joseph A. Pratt, "The Ascent of Oil: The Transition from Coal to Oil in Early Twentieth-Century America," in *Energy Transitions: Long-Term Perspectives*, ed. Lewis J. Perelman (Boulder, Colo.: Westview Press, 1981). Mark H. Rose has produced a number of useful works on energy, particularly natural gas, including *Cities of Light and Heat: Domesticating Gas and Electricity in Urban America* (University Park: Pennsylvania State University Press, 1995); "There Is Less Smoke in the District: J. C. Nichols, Urban Change, and Technological Systems," *Journal of the West* 25, no. 2 (1986): 44–54; and with John G. Clark, "Light, Heat, and Power: Energy Choices in Kansas City, Wichita, and Denver, 1900–1935," *Journal of Urban History* 5 (1979): 340–60. Also on natural gas see John H. Herbert, *Clean Cheap Heat: The Development of Residential Markets for Natural Gas in the United States* (New York: Praeger, 1992).

For background on the coal industry I referred to James P. Johnson, *The Politics of*

Soft Coal: The Bituminous Industry from World War I through the New Deal (Urbana: University of Illinois Press, 1979). Particularly for wartime coal policies see John G. Clark, *Energy and the Federal Government: Fossil Fuel Policies, 1900–1946* (Urbana: University of Illinois Press, 1987). Alfred D. Chandler may have exaggerated the importance of anthracite coal to the genesis of industrialization in the United States, but his article on that fuel provides a fine narrative of the early coal trade. See "Anthracite Coal and the Beginnings of the Industrial Revolution in the United States," *Business History Review* 46 (1972): 141–81. For the lives of coal miners and the importance of the fuel to the mining region see David Alan Corbin, *Life, Work, and Rebellion in the Coal Fields: The Southern West Virginia Miners, 1880–1922* (Urbana: University of Illinois Press, 1981). I also referred to Robert J. Cornell, *The Anthracite Coal Strike of 1902* (New York: Russell & Russell, 1957). For a more contemporary and romantic source on coal see Robert W. Bruere, *The Coming of Coal* (New York: Association Press, 1922). Less romantic is William Jasper Nicolls, *The Story of American Coals* (Philadelphia: J. B. Lippincott, 1897).

On electricity see Harold Platt, *The Electric City: Energy and the Growth of the Chicago Area, 1880–1930* (Chicago: University of Chicago Press, 1991). I also relied heavily on the several works available on railroad electrification, particularly Carl Condit's *The Port of New York: A History of the Rail and Terminal System from the Beginnings to Pennsylvania Station* (Chicago: University of Chicago Press, 1980). More valuable for the Pennsylvania Railroad specifically is Michael Bezilla, *Electric Traction on the Pennsylvania Railroad, 1895–1914* (University Park: Pennsylvania State University Press, 1980).

Engineers

Stanley K. Schultz's *Constructing Urban Culture: American Cities and City Planning, 1800–1920* (Philadelphia: Temple University Press, 1989) has been immensely important to this study, particularly in its discussion of municipal engineers, but also in its portrayal of late-nineteenth-century cities generally. Also of value is his article with Clay McShane, "To Engineer the Metropolis: Sewers, Sanitation, and City Planning in Late-Nineteenth-Century America," *Journal of American History* 65 (1978): 389–411. Anyone interested in the history of the engineering profession should start with Edwin T. Layton Jr., *The Revolt of the Engineers: Social Responsibility and the American Engineering Profession* (Cleveland: Press of Case Western Reserve University, 1971). Of less value is David F. Noble's *America by Design: Science, Technology, and the Rise of Corporate Capitalism* (New York: Knopf, 1977). Noble's radical agenda leads him to rather dubious conclusions, including the assertion that engineers served only the dominant class of society. On scientific management and the efficiency craze of the 1910s, see Samuel Haber, *Efficiency and Uplift: Scientific Management in the Progressive Era, 1890–1920* (Chicago: University of Chicago Press, 1964). Although not specifically about engineers, Thomas G. Marx's "Technological Change and the Theory of the Firm: The American Locomotive Industry, 1920–1955," *Business History Review* 50 (1976): 1–24, provides insight into the role of engineering developments within railroad firms. Marx's article also gives fine detail concerning the railroad industry's shift from coal to diesel.

Environmental History

In preparing this work I was struck at the glaring absence of a complete history of conservationism. Perhaps stating that Samuel Hays's ground-breaking work, *Conservation and the Gospel of Efficiency: The Progressive Conservation Movement, 1890–1920* (Cambridge: Harvard University Press, 1959), is still the most valuable single work on the topic best illustrates the need for a fresh synthetic work. Also of value, however, are Stephen Fox, *The American Conservation Movement: John Muir and His Legacy* (Madison: University of Wisconsin Press, 1991); John F. Reiger, *American Sportsmen and the Origins of Conservation* (Norman: University of Oklahoma Press, 1986); and perhaps most important, Gifford Pinchot, *The Fight for Conservation* (New York: Doubleday, Page, 1910).

On early environmental activism see the essays in Martin Melosi's *Pollution and Reform in American Cities, 1870–1930* (Austin: University of Texas Press, 1980). Melosi's *Garbage in the Cities: Refuse, Reform, and the Environment, 1880–1980* (College Station: Texas A & M University Press, 1981) is also useful. More recently, Robert Gottlieb argued that the roots of modern environmentalism lay in progressive reform, a position I clearly accept, but he failed to support his argument adequately. Nonetheless, *Forcing the Spring: The Transformation of the American Environmental Movement* (Washington: Island Press, 1993) contains valuable analysis. Although largely outside my chronology, Andrew Hurley's *Environmental Inequalities: Class, Race, and Industrial Pollution in Gary, Indiana, 1945–1980* (Chapel Hill: University of North Carolina Press, 1995) offers a fine history of urban environmental activism and governmental response. Also of interest is Hurley's "Creating Ecological Wastelands: Oil Pollution in New York City, 1870–1900," *Journal of Urban History* 20 (1994): 340–64; and Craig E. Colton, "Creating a Toxic Landscape: Chemical Waste Disposal Policy and Practice, 1900–1960," *Environmental History Review* 18 (1994): 85–116.

William Cronon's *Nature's Metropolis: Chicago and the Great West* (New York: W. W. Norton, 1991) has deeply influenced my thinking about urban and environmental history. With its argument concerning the connections between city and country and its rich detail of Chicago in the 1800s, this work proved to be among the most important monographs I read. The impetus for my study largely came as a reaction to Samuel Hays, *Beauty, Health, and Permanence: Environmental Politics in the United States, 1955–1895* (New York: Cambridge University Press, 1987). Here Hays describes an environmental ethic centered on the search for environmental amenities as developing in the affluent 1950s, an assertion I could not trust.

For a fine discussion of urbanites' thoughts on open space and recreation see David

Schuyler, *The New Urban Landscape: The Redefinition of City Form in Nineteenth-Century America* (Baltimore: Johns Hopkins University Press, 1986). William H. Wilson's work on City Beautiful is thorough and convincing. See *The City Beautiful Movement* (Baltimore: Johns Hopkins University Press, 1989). Also of value is Jon A. Peterson, "The City Beautiful Movement: Forgotten Origins and Lost Meanings," *Journal of Urban History* 2 (1976): 415–34. On the construction around the newly electrified Grand Central see Kurt C. Schlichting, "Grand Central Terminal and the City Beautiful in New York," *Journal of Urban History* 22 (1996): 332–49.

Urban history

As this study is ultimately a work about American cities, I have found the work of several urban historians very useful. Paul Boyer's *Urban Masses and Moral Order in America, 1820–1920* (Cambridge: Harvard University Press, 1978) greatly influenced my thinking on progressive reform, particularly in establishing reformist roots in Victorian ideals concerning morality, health, and cleanliness. I found Jon C. Teaford's arguments concerning the success of municipal government largely convincing. See *The Unheralded Triumph: City Government in America, 1870–1900* (Baltimore: Johns Hopkins University Press, 1984). Like most urban historians I have been influenced by the work of Kenneth Jackson, particularly *The Crabgrass Frontier: The Suburbanization of the United States* (New York: Oxford University Press, 1985). I relied on several works for the histories of the individual cities studied in this work. Among the most helpful were David W. Lewis, *Sloss Furnaces and the Rise of the Birmingham District: An Industrial Epic* (Tuscaloosa: University of Alabama Press, 1994); Roy Lubove, *Twentieth Century Pittsburgh: Government, Business, and Environmental Change* (New York: John Wiley, 1969); Zane Miller, *Boss Cox's Cincinnati: Urban Politics in the Progressive Era* (Chicago: University of Chicago Press, 1968); Steven J. Ross, *Workers on the Edge: Work, Leisure, and Politics in Industrializing Cincinnati, 1788–1890* (New York: Columbia University Press, 1985).

My work has also been influenced by several other urban history monographs, including Thomas Bender, *Toward an Urban Vision: Ideas and Institutions in Nineteenth-Century America* (Baltimore: Johns Hopkins University Press, 1975); Lizabeth Cohen, *Making a New Deal: Industrial Workers in Chicago, 1919–1939* (New York: Cambridge University Press, 1990); David M. Emmons, *The Butte Irish: Class and Ethnicity in an American Mining Town, 1875–1925* (Urbana: University of Illinois Press, 1989); and Sam Bass Warner Jr., *The Urban Wilderness: A History of the American City* (New York: Harper & Row, 1972).

Progressivism

Several works concerning progressive reform have shaped my thinking on the era from the late 1800s through 1920. I have been deeply influenced by Samuel Hays and the organizational model outlined in *The Response to Industrialism, 1885–*

1914 (Chicago: University of Chicago Press, 1957). Robert Wiebe's *The Search for Order, 1877–1920* (New York: Hill & Wang, 1967) still offers the finest portrait of progressivism available. More recent works, including Nell Irvin Painter, *Standing at Armageddon: The United States, 1877–1919* (New York: W. W. Norton, 1987), are useful in adding the stories of those left out of earlier narratives. Although Gabriel Kolko's *The Triumph of Conservatism* (New York: Free Press, 1963) reads like historical hyperbole, his main argument concerning the role of business in shaping reform has influenced my work.

Students of progressivism will also find much of value in several articles, many of which are now fairly old. See Samuel Hays, "The Politics of Reform in Municipal Government in the Progressive Era," *Pacific Northwest Quarterly* 55 (1964): 157–69; John Higham, "The Reorientation of American Culture in the 1890s," in *The Origins of Modern Consciousness,* ed. John Weiss (Detroit: Wayne State University Press, 1965); Arthur Link, "What Happened to the Progressive Movement in the 1920s?" *American Historical Review* 64 (1959): 833–51; and Daniel Rodgers, "In Search of Progressivism," *Reviews in American History* 10 (1982): 113–32.

For a discussion of women and politics before winning the franchise, see Paula Baker's important article, "The Domestication of Politics: Women and American Political Society, 1780–1920," *American Historical Review* 89 (1984): 620–47. Maureen A. Flanagan's recent work also provided important context to my work on women reformers. See "Gender and Urban Political Reform: The City Club and the Woman's City Club of Chicago in the Progressive Era," *American Historical Review* 95 (1990): 1032–50; and "The City Profitable, The City Livable: Environmental Policy, Gender, and Power in Chicago in the 1910s," *Journal of Urban History* 22 (1996): 163–90. Mary Ritter Beard's early work on women and reform is remarkably insightful. See *Woman's Work in Municipalities* (New York: D. Appleton, 1915). See also Marlene Stein Wortman, "Domesticating the Nineteenth-Century American City," *Prospects* 3 (1977): 531–72. On women's social organizations see Karen J. Blair, *The Clubwoman as Feminist: Womanhood Redefined, 1868–1914* (New York: Holmes & Meier, 1980).

Index

Page numbers in *italics* denote illustrations. Those in **boldface** denote tables.

Library of Congress Cataloging-in-Publication Data

Stradling, David
  Smokestacks and progressives : environmentalists,
engineers, and air quality in America, 1881–1951 / David
Stradling.
     p. cm.
  Includes bibliographical references and index.
  ISBN 0-8018-6083-0 (alk. paper)
    1. Air—Pollution—United States—History.   2. Smoke
prevention—United States—History.   I. Title.
TD883.2.S77   1999
363.739'2'0973   dc21   99-11410
                                                              CIP

Wartime & efficiency   vs economic expansion

Shifting focus of abatement – pg 151
                                                pg 149

  Roller coaster of abatement –
      stopped in time of nat'l crises – WW1,
                                          Great Depr
  increasingly _technical_ mvmt → like industry
                                            itself
                    155

Printed in the United States
86009LV00003B/91-138/A

9 780801 872501